NASA

V·I·S·I·O·N·S O·F S·P·A·C·E

Capturing the History of NASA

ROBIN KERROD

COURAGE
BOOKS

Canadian representatives: General Publishing Co., Ltd.,
30 Lesmill Road, Don Mills, Ontario M3B 2T6.

9 8 7 6 5 4 3 2 1

Digit on the right indicates the number of this printing.

Library of Congress Cataloging-in-Publication Number
  90-81546
  ISBN 0-89471-853-3

This book was devised by John Strange and Robin Kerrod, and
produced by Multimedia Books Limited,
32-34 Gordon House Road, London NW5 1LP.

Printed in Italy by Imago. All rights reserved under the
Pan-American and International Copyright conventions.

This book may be ordered by mail from the publisher.
Please add $2.50 postage and handling for each copy.
But try your bookstore first!

Published by Courage Books, an imprint of
Running Press Book Publishers,
125 South Twenty-second Street,
Philadelphia, Pennsylvania 19103.

Typeset by O'Reilly Clark Typesetting, London.

C·O·N·T·E·N·T·S

1 Through an artist's eyes

"In the long run, the truth seen by an artist is more meaningful than any other type of record. . . . Future generations will realize that we had not only the scientists and engineers capable of shaping the destiny of our age, but also artists worthy to keep them company."

Dr. H. Lester Cooke, former Curator of Painting at the National Gallery of Art and principal art advisor to NASA, 1963–1972.

The conquest of space has arguably been the supreme achievement of technological man. It is astonishing to think that a century ago the first airplane had not yet flown. Yet today we can jet around the world in a matter of hours, send people to the Moon, and dispatch our eyes and ears on robot probes to other worlds so far away that it takes beams of light hours to reach them. As the *Washington Post* commented on the occasion of the first lunar landing in July, 1969: "The creature who had once stood blinking at the door of his paleolithic cave has come a long way. No longer is he tied to the area or even the world where he was born. The heavens lie open now."

The Space Age was scarcely a year old in October, 1958, when NASA, America's National Aeronautics and Space Administration, came into being. Since then it has guided the U.S. space program through fair times and foul, often setting the pace for other nations to follow. And it has charted the painstaking progress into the alien world of space — from launch pad to orbit, to Moon landing, to Shuttle, to space station — faithfully and from every conceivable angle through the medium of photography. Yet in these often stunning "drawings with light", there seems something intangibly lacking. That something is the awe, the emotion, the spirit, the imaginative interpretation that such epoch-making events as the first human footprint in the lunar soil should inspire in us mortals, shackled to Earth by gravity since the dawn of time.

It quickly became clear that to document the American thrust into the universe through multidimensional perspectives, an artist's eye was required. This led the second NASA Administrator, James E. Webb, in 1962 to institute the NASA Art Program in order, he said, "to record the spirit as well as the sights of the Space Age." Since that time, NASA has invited numerous artists to make their own record of space launches and the many other aspects of this most exciting phase of human endeavor.

There are many other facets to NASA's art: artists' impressions, necessary in the visualization of future hardware and mission operations; mission emblems, designed to personalize and reflect the content or spirit of a mission; and computer wizardry, which can, for example, take electronic gossamer from distant space probes and convert it into a riveting kaleidoscope of superdistant worlds.

Artist and easel
Attila Hejja at work with a traditional easel on August 29, 1983, producing a painting of orbiter *Challenger* on Launch Pad 39A. At about 2:30 A.M. next day *Challenger* makes a spectacular night launch (STS-8), the first ever by the Shuttle.

In welcome shade
Sheltering from the hot spring sunshine in March, 1982, Robert McCall completes two small watercolor sketches of *Columbia*, awaiting its third flight into space (STS-3).

Giving a unique insight
The desire to "record the spirit as well as the sights" of momentous events dates back in the United States to the days of the War of Independence, when an artist served on the staff of General George Washington at the Valley Forge encampment in Pennsylvania. Artists were also on hand to record the pushing back of the "Wild West" frontier, the march of the railroad, and the strife and slaughter of the Civil War.

So there were plenty of precedents to encourage NASA Administrator James Webb to initiate a NASA Art Program in 1962. Encouragement came from top-level NASA management, and expert advice came from Dr. H. Lester Cooke, then Curator of Painting at the National Gallery of Art in Washington, D.C. Cooke remained principal art advisor to NASA until his death in 1973. During this time the program was directed by James D. Dean, latterly Curator of Art at the National Air and Space Museum, Washington.

The NASA Art Program had its practical beginning in the spring of 1963. In a press release announcing the program, Administrator Webb said: "Important events can be interpreted by artists to give a unique insight into significant aspects of our history-making advance into space. An artistic record of this nation's program of space exploration will have great value for future generations and may make a significant contribution to the history of American art."

At Cape Canaveral
The Art Program was up and running just in time for the final mission in the Mercury project, which pioneered U.S. manned space flight. This was the mission in May, 1963, in which Gordon Cooper in the Mercury capsule *Faith 7* was aiming to break the U.S. space endurance record with a 22-orbit flight.

A team of seven artists was invited to work on location around the launch site at Cape Canaveral; one was also asked to travel on the aircraft carrier that would recover the capsule after splashdown in the Pacific Ocean. The artists covered the prelaunch activities side by side with the media newsmen and women. With makeshift easels and sketch pads they roamed around the Cape, which bristled, then more than now, with stark gantries, painted a beautiful red as protection against the ravages of the salt-laden air.

Such objects were of extreme fascination for the "original seven" artists. Wrote one of them, Peter Hurd: "The cranes are of open steelwork, an interlacing maze of girders and tubing, lavishly lighted from inside and out, giving an unbelievably realistic effect of incandescent filigree. Incongrous and wonderful as this was, it was only the beginning."

Since those first artists put pen and brush to paper and canvas at the Cape, every major launch has seen artists gather around launch pads and landing sites. They have talked to the astronauts and flown in simulators on imaginary missions into orbit and beyond.

Tools of the trade
On his knees and with the tools of his trade about him, Paul Arlt works on a painting of Launch Pad 39B in April, 1983, at the time of *Challenger*'s maiden flight into orbit.

Kennedy Spaceport
This aerial view of the Kennedy Space Center shows the towering Vehicle Assembly Building (VAB) and beyond it the Shuttle landing strip. Note the rocket on its side in front of the VAB. This is a Saturn V, the 36-story high Moon rocket for which the VAB was built.

"Gantry Structures"
Since the beginning of the NASA Art Program, artists have been fascinated by the crisscross girder work of the towering launch-gantry structures. This interpretation was painted by Lamar Dodd.

"Gantry at Night"
As the countdown to a launch ticks away, the work continues on the launch pad day and night. One of the original seven artists invited to participate in the NASA Art Program, Peter Hurd, painted this night scene, in which the gantry becomes "incandescent filigree".

Going operational
A close-up of Alfred McAdams' painting of *Columbia* on the launch pad, executed in November, 1982, before it lifted off on its fifth mission (STS-5), the first operational flight of the Shuttle.

Sketches
Bill Phillips prepares a series of small sketches of *Columbia* on the launch pad in November, 1982, on which he will later base a large studio painting.

Adding a new dimension
Well over 100 artists have participated in the NASA Art Program since its inception. They represent the whole spectrum of artistic styles — from traditionalists such as Paul Calle and Robert McCall to modernists such as Robert Rauschenberg and Lamar Dodd — and the complete range of mediums and techniques — pen and ink, charcoal, oils, watercolors, acrylics, silk screen, color Xeroxes, and so forth.

NASA has never told an artist what to draw or paint. As James Dean recalled: "On the contrary, he was told that NASA was commissioning his imagination first and his technical skill second. He was completely free to interpret the subject in any way he wished. NASA simply wished to add a new dimension to the understanding of an epic page in history."

Launch Pad 39B
An aerial view of Launch Pad 39B at the Kennedy Space Center. Like its twin 39A, it is located close to the shore at Cape Canaveral. Here orbiter *Atlantis* is being readied for its third flight into orbit (STS-27) in November, 1989.

The results of NASA's commissions have vindicated Webb's vision in setting up the Program. Observed Frank Getlein, respected art critic of the *Washington Star* in 1971: "What they [NASA] have created is an extraordinary record because it records more than the eye, even the trained eye, can see. The sketches and many of the paintings do indeed record what any eye can see, but the seeing is special. Beyond that, artist after artist at some point found himself pushed by the material itself into taking an abstract, symbolic view of the space flights themselves, painting a vision, not a view."

Into the Shuttle era
Currently the Art Program is the most vibrant it has ever been. It was run down in the early 1970s following the completion of the Apollo project and the spin-off Skylab missions. But in 1977, with the imminent arrival of the Space Shuttle, it was revived and revitalized. Since that time it has been under the enthusiastic direction of Robert Schulman, who, like former director James Dean, is himself an accomplished artist and graphic designer.

Let us leave it to the celebrated science-fiction writer Ray Bradbury to sum up the importance of the Art Program. It is, he says, "a shining illustration of the metaphor of NASA, capturing symbolically the essence of human inspiration, which lies as much at the heart of a billion-dollar spacecraft as it does at the heart of the world's most sublime artistic expressions."

Pencil drawing
Pencil and pad are Nixon Galloway's preferred medium as he works on a drawing of *Challenger* in August, 1983, prior to its third flight into space (STS-8).

"Time, Space and *Columbia*"
The press site at Complex 39 of the Kennedy Space Center is the hub of media activity during a Shuttle launch. It is located near the VAB and overlooks Launch Pad 39A across one of the many lagoons that dot the landscape. This painting by Billy Morrow Jackson records the scene at the press site during STS-2, the second flight of the Shuttle in November, 1981. The artist is shown in black cap and red shirt in the foreground, surveying a panorama of Shuttle activities. The painting depicts a variety of Shuttle events, occurring at different times, in order to tell the entire story of the lift-off.

"One Flight Out, One Flight In"
The Kennedy Space Center is
not as disruptive to the
environment as might be
thought. It is located in the
middle of a wildlife haven —
the Merritt Island Wildlife
Refuge. In this refuge there are
some 500 species of bird, fish,
mammal, amphibian, and
reptile. Several eagles nest at
the Center, which inspired
Robert Burnell to produce this
painting depicting the maiden
launch of *Challenger* (STS-6) in
April, 1983.

"Florida Coast — Fire Pillar"
The first launch of *Columbia*, on April 12, 1981, inspired James Cunningham to produce this painting. It concentrates on the tall column of exhaust matter that issues from the solid rocket boosters.

Different interpretations

If you ask a dozen witnesses of, say, a bank robbery, to describe what took place, the chances are that you will get a dozen different answers. The thieves will be short, fat, and fair, but also tall, lean and dark! Their getaway vehicle will be a blue Chevvy, a green Ford, and so on.

People interpret what they see in a variety of ways. So it is with an artist when he or she is asked to create a record, an impression, an interpretation of a fleeting moment in time. The greater the inspiration, the greater the impetus to create something inimitable, unique. There is infinite inspiration in the human thrust into space. In, for example, the dazzling spectacle of a rocket launch, with the unleashing of such blazing fire, such thundrous noise, such awesome power. And in what it all means — that we are poised on the threshold of a glorious new age, in which the children of our children's children may be born into a world that is not Earth. That we have climbed the first steps on the stairway that may one day take us to the stars.

It is fascinating to see how different artists react to the same inspiration. Three examples of artistic variation on a space theme are shown here, the theme being the launch of the Space Shuttle. The strongly colored launch-pad picture is an artist's impression of a Shuttle lift-off, deliberately intended to be true to life. But it was very much a product of the artist's imagination since it was painted five years before the first Shuttle took off.

Shuttle lift-off
With its three main engines and twin solid rocket boosters blazing, the Space Shuttle lifts off the launch pad. Together, the five rockets deliver over 7 million pounds (3 million kilograms) of thrust. The twin boosters and external fuel tank will fall away as the winged orbiter makes its journey into space.

"The Second Great Step"
The first giant step into the universe occurred when Apollo astronaut Neil Armstrong planted his footprints in the lunar soil in July, 1969. The first flight of the Space Shuttle, with orbiter *Columbia*, is arguably the second. So thought Lamar Dodd when he produced this painting.

The hardware is rendered more or less faithfully, but there are details that we can now identify, with hindsight, as being inaccurate. For example, the exhaust gases issuing from the main engines are in reality nearly invisible. There is much less flame during a Shuttle lift-off but a great deal more steam. This is caused by water being dumped on the launch platform to reduce heat damage to the Shuttle's insulating tiles. And so on. Nevertheless, the picture gives a good idea of what the real lift-off would look like, as was intended.

The other two interpretations of a Shuttle launch are paintings executed on the occasion of the maiden flight of the Shuttle, when orbiter *Columbia* reached for the skies for the first time. In "Florida Coast — Fire Pillar" the artist focuses his attention on the great column of flame and smoke that spews from the nozzles of the twin solid rocket boosters. The column eventually begins to totter as the winds at various levels cut into it. Blues are dominant, as one would expect with the deep blue Florida sky and azure Atlantic as a backcloth. There is a hint of a reflection in the foreground in one of the lagoons that dot the landscape at Cape Canaveral.

"The Second Great Step" sees *Columbia* blasting away from the eruption of fire, steam, and smoke on the launch pad — or does it? The artist, Lamar Dodd, would not necessarily agree or disagree with this interpretation. He feels that viewers should see what they want to see. "A lot of dreaming and planning goes into a canvas," he says. "I envision certain things that could happen, some that do and possibly some that don't. In the final analysis I think the artist should have his freedom to create. So if the viewer looks at this canvas and wonders about the lavender area at the left, maybe it is the shoreline. To me it's just beautiful color and a wonderful pattern. If the big white area is a white poppy, that's fine. It could be a cloud. It could be some thrust from the ship itself combined with the cloud formation. But to me it's just a gorgeous, beautiful shape that flowers like a poppy, which represents life itself."

Go, *Discovery*!
An ultrawide-angle lens
squeezes in much of the
launch pad area as it records
the spectacular lift-off of
Discovery on Shuttle mission
51-C in January, 1985.

Through the camera lens

It is interesting to compare the very different artists'
interpretations of a Shuttle launch shown on the
previous pages with a fourth, here viewed through a
camera lens, a special camera with a special lens,
mounted in a special location, in a shockproof "firebox"
255 feet (78 meters) up the launch-pad service
structure. The result is a unique perspective of the rising
Shuttle and the launch-pad area.

The camera is an IMAX movie camera, which uses
special 70 mm filmstock, with 10 times the area of
standard 35 mm film and three times the area of standard
70 mm film. This allows the film to be projected in
IMAX cinemas on to giant screens up to seven stories
high and 100 feet (30 meters) wide.

The picture is a still taken from the IMAX movie
"The Dream Is Alive", which was premiered in 1985
and has now been seen by more than 15 million people.
Veteran Shuttle astronaut Robert Crippen rates
watching the movie as "the closest thing there is on
Earth to being in space".

"The Dream is Alive" and its sequel "The Fragile
Earth" reveal the beauty of our planet to the eyes of
ordinary people who cannot ride the Shuttle into orbit.
Such beauty was first appreciated by astronaut Gordon
Cooper, who flew the final mission in the Mercury
project, pioneering manned flight for the U.S. In his day
and a half in space he had more time than astronauts
before him to sit back and admire the view.

Cooper's report of what he saw in orbit created a
sensation. Mission scientists thought he must have
suffered from some form of space-induced visual
hallucinations. "I could detect," said Cooper,
"individual houses and streets in the low humidity and
cloudless areas such as the Himalayan mountain area,
the Tibetan plain, and the southwestern desert area of
the U.S. ... I saw what I took to be a vehicle along a road
... in the Arizona–West Texas area. I could see the dust
blowing off the road, then could see the road clearly,
and when the light was right, an object that was probably
a vehicle."

All this he saw from an altitude of between 100 and 160 miles (160 and 260 kilometers). No wonder mission scientists were incredulous!

When the longer Gemini flights got underway and there were two pairs of eyes to look down from orbit, the truth of Cooper's observations was confirmed — incredible detail can be seen from orbit. Since then photography has played a major role during space missions in Earth orbit and particularly during the Moon-landing flights of Apollo. What vistas ravished the eyes of the two dozen astronauts who planted their footprints in the lunar soil. And what photographers they turned out to be, capturing for us with their lenses the haunting beauty of the desolate lunar landscapes. Even today, decades afterwards, their photographs still stun the eye and excite the imagination.

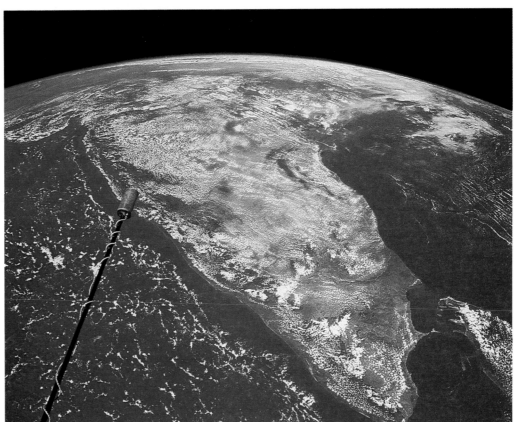

Magnificent desolation
At the *Apollo 17* landing site at Taurus-Littrow, geologist Harrison Schmitt removes tools from the lunar roving vehicle. The photograph was taken by fellow moonwalker Eugene Cernan, in December, 1972.

Gemini spectacular
The *Gemini 9* astronauts photograph this spectacular view of India in June, 1966. It is one of the finest of many fine Earth photographs taken in orbit on the Gemini missions.

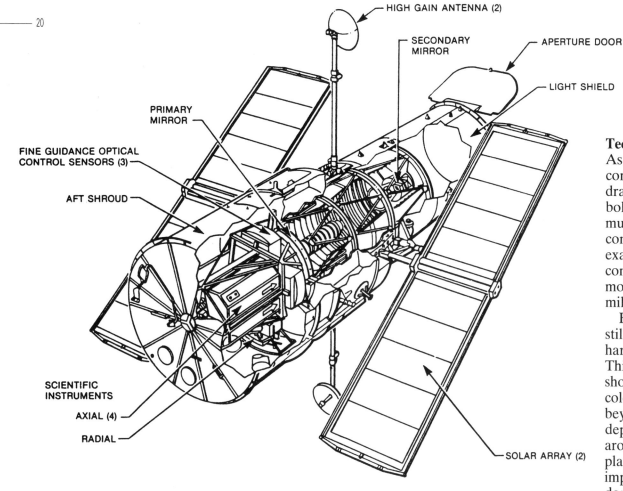

HIGH GAIN ANTENNA (2)

SECONDARY MIRROR

APERTURE DOOR

LIGHT SHIELD

PRIMARY MIRROR

FINE GUIDANCE OPTICAL CONTROL SENSORS (3)

AFT SHROUD

SCIENTIFIC INSTRUMENTS

AXIAL (4)

RADIAL

SOLAR ARRAY (2)

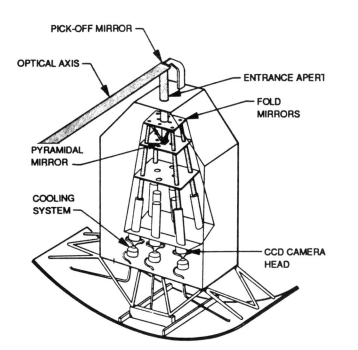

PICK-OFF MIRROR

OPTICAL AXIS

ENTRANCE APERT

FOLD MIRRORS

PYRAMIDAL MIRROR

COOLING SYSTEM

CCD CAMERA HEAD

Line cutaway
A line drawing may contain all the necessary information to assist an explanation about how something works. But it is difficult in just black and white to work out exactly what is going on.

Technical drawings
As in any engineering enterprise, the design and construction of space hardware requires the most basic drawings at the drafting level to show how the nuts and bolts of various systems fit together. The scale of the mundane draftsmanship involved in the design and construction of the Apollo/Saturn launch vehicles, for example, can hardly be imagined — two million components went into the Apollo command and service modules, one million into the lunar module, and three million into the 36-story high Saturn V rocket!

Even in this technologically sophisticated age there is still a need for artwork that demonstrates how space hardware and systems dovetail together and operate. This is often accomplished with simple line artwork, showing cutaway views. This can then be enhanced with color for publicity leaflets, magazines, and books. Going beyond this, the artist is invited to use his imagination to depict how the hardware will appear in situ — in orbit around the Earth, on the Moon, en route to the outer planets, and so forth. This results in the realistic artist's impression or concept. It is in this field as well as in the domain of fine art that NASA has produced such artistic wealth.

Here the subject of the Hubble Space Telescope is chosen to illustrate different levels of technical art, from cutaway line artwork to full-blown artist's impression. The line drawings are taken from the Shuttle mission STS-31 press kit of April, 1990.

The Hubble Space Telescope is a particularly exciting piece of hardware, as tall as a house and with an Earth weight of 11 tons. Developed at a suitably astronomical cost of $1.5 billion, it should see into the depths of space with ten times the clarity of Earth-based telescopes, which are plagued by atmospheric shimmer.

The Space Telescope was ready for launch in 1986, but the demise of *Challenger* and the consequent grounding of the Shuttle delayed its launch until 1990. The delay gave the design team the opportunity of installing the latest electronics, which will further upgrade the images it will send back. If the Hubble Space Telescope lives up to its promise, it will bring about a revolution in astronomical viewing as great as that achieved by Galileo when he first trained a telescope on the heavens nearly four centuries ago.

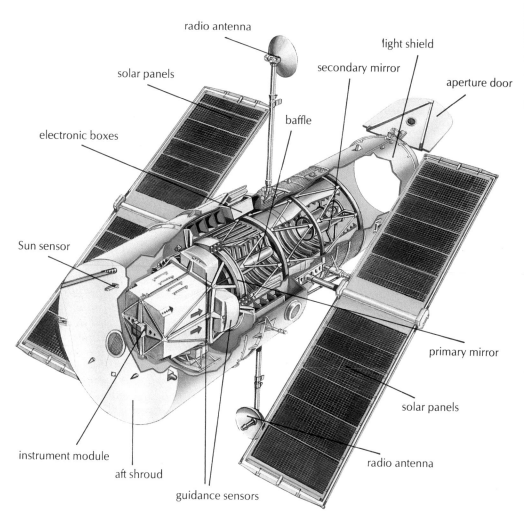

labels for image 1:
radio antenna
solar panels
electronic boxes
baffle
secondary mirror
light shield
aperture door
Sun sensor
primary mirror
instrument module
solar panels
aft shroud
guidance sensors
radio antenna

Color cutaway
When the artist gets to work with color, the flat piece of art takes on added dimensions. The different systems can be traced at a glance. Here, in the Hubble Space Telescope, light from the stars enters through the aperture door and falls on the primary mirror, which measures 94 inches (2.4 meters) across. This reflects the light onto the secondary mirror, which in turn reflects it into the baffle. The light is brought into focus within the instrument module, where electronic systems record the image.

Telescope in space
Up in orbit some 350 miles (550 kilometers) above the Earth, the Hubble Space Telescope looks deep into the universe. In this artist's impression, a Shuttle orbiter is in attendance. It is maneuvering into position to carry out the regular in-orbit servicing of the telescope.

▲
"One Man's Lifetime"
This artwork by Roland O. Powell depicts the progress that has taken place over a single human lifespan, from the flightless days of the horse and buggy era to the present era, when the Space Shuttle commutes regularly into orbit.

"Flight out of Time"
Susan Kaprov was inspired to create this work by the expansion into space that the Shuttle promised.

Grand themes

As well as portraying individual events, operations, and pieces of hardware, artists have also been inspired to more visionary renderings of what might be called the "grand themes" of the space age: the ascent of humankind into the cosmos; our relationship to the universe; the space-time relationship; and so forth. Three quite different themes and artistic approaches are illustrated here.

"One Man's Lifetime" uses what might be called the "documentary" approach to illustrate how the conquest of space has occurred within the span of a human lifetime. First comes the horse and buggy, then the conquest of the air with biplane and jet, then the leap into orbit, and finally the touch-down on the Moon.

"Flight out of Time" is a symbolic interpretation of space and the Shuttle era, which uses the medium of color Xerox prints rather than painting. It is a fascinating collage of images that rewards lengthy study. Birds' wings flutter over nearly every segment, here with a Lunar Orbiter image of the Moon, there with a total eclipse of the Sun, and there again with a pristine Shuttle orbiter poised for lift-off. Eclipses are a recurrent theme, here showing the flaming corona at totality, there revealing the diamond-ring effect immediately afterwards.

"Spirit of Man in Space", dating from the Apollo era, is an example of an artist's personal insight into a grand theme that may be interpreted by others in a variety of ways. It is not immediately clear even which way up it should be! Here is one interpretation, which may differ from the artist's. But that should not matter. Let the viewer interpret the art in the way he or she wishes.

The would-be astronauts lie flat on their backs as they must in order to withstand the g-forces created by the rocket that blasts them into the heavens. Their partial nakedness signifies their fragility in the alien environment into which they ascend. A hand reaches out to touch the globe of Earth, with its blue annulus of atmosphere. In the distance at left is the Vehicle Assembly Building, from where the launch rocket began its journey into space. At right is the rocket punching its way through the atmosphere at many times the speed of sound to reach the airless void where the Sun never sets.

"Spirit of Man in Space"
The first manned flight in the Apollo program, *Apollo 7*, in October, 1968, prompted Tom O'Hara to execute this symbolic work. This flight marks the beginning of one of the most exciting series of adventures since the dawn of civilization.

Apollo 12 patch
Astronauts Charles Conrad,
Alan Bean and Richard
Gordon wear this crew badge
when they lift off amid
thunderbolts on November
14, 1969, to make the second
lunar landing. The theme of the
sailing clipper is continued in
the callsign for the CSM,
Yankee Clipper.

Space-patch art

On every U.S. manned space flight since *Gemini* 5 in 1965, astronauts have worn a mission patch, or emblem, on their flight suit, and often on their in-flight casual wear as well. They wear the patch on the right breast. On the left breast they wear the traditional NASA extended vector logo, widely referred to as the "meatball".

The practice of wearing mission patches originated in the desire by the astronauts to personalize their flights. And from the start the astronauts themselves have provided the basic theme and ideas for the patches, which artists have then translated into finished designs.

Over the years space-patch design has developed into an exciting art form in its own right. It has come into its own particularly in the Shuttle era as the pace of launches has accelerated. Patches also play a prominent role on Shuttle memorabilia and souvenirs, from mugs to T-shirts.

Early in the Shuttle program NASA almost dropped the idea of a new crew patch for each flight, because it felt that the patches and spin-off items would not be designed and distributed fast enough between increasingly frequent missions. Public and astronaut opinion, however, demanded that the crew patches stay.

The patch concept has also extended to include emblems for unmanned space programs such as Viking and Voyager; for hardware such as the MMU (manned maneuvering unit) and Hubble Space Telescope; and for commemoratives such as NASA's first 25 years and the 20th anniversary of the first landing on the Moon. All have now become attractive "collectables" for a great many people worldwide.

There is considerable artistry in the execution of the patches, which are Swiss-embroidered, that is precision-embroidered by machine to the most exacting standards. Patches are always fully embroidered, with all of the backing fabric covered. Lustrous rayon yarns in up to 19 colors are built up on the fabric to create a dimension lacking in the flat art.

Teacher in Space

NASA

Teacher's flight
This symbolic logo was designed by the current director of the NASA Art Program, Robert Schulman. On the first flight in the Teacher in Space project Christa McAuliffe was to have given the first lesson from space to the nation's schoolchildren. But that was the fateful *Challenger* flight of January 20, 1986, which ended in disaster 73 seconds after lift-off.

Shuttle 41-B patch
On Shuttle mission 41-B, orbiter *Challenger* carries a crew of five and has three main objectives, reflected in the design of the patch. They are to launch two satellites, Westar VI and Palapa B, and thoroughly test the MMU (manned maneuvering unit). Both satellites are launched, but both become stranded in low orbit. The tests of the MMU by Bruce McCandless and Robert Stewart, however, are spectacularly successful.

Three examples of emblem designs are shown here: the crew patches of *Apollo 12* and Shuttle mission 41-B, and the logo for the Teacher in Space program. There is a story behind each one.

The *Apollo 12* patch was worn by the crew who made the second lunar landing. The clipper ship gives a hint that the crew members were all drawn from the U.S. Navy and symbolically relates the era of the fast sailing clipper to the era of space flight. Just as clippers brought foreign shores closer to the U.S. and mediated increased utilization of the seas, so spacecraft opened the way to other planets and *Apollo 12* helped to bring about increased utilization of space. The portion of the Moon illustrated shows the Ocean of Storms where *Apollo 12* landed.

The design of the Shuttle 41-B patch has a more practical theme, reflecting events scheduled to take place on the mission. The orbiter *Challenger*, wheels down for landing, dominates the design. It is flanked by scenes showing two main mission operations, the launch of a satellite with a PAM (payload assist module) booster, and the first test-flight of the MMU.

The Teacher in Space logo is much more symbolic and particularly striking in its use of shape and color. A Shuttle launch is superimposed on a flaming torch, with the blue Earth in the background, set against a star-studded, deep velvet backcloth of interstellar space. The flaming torch symbolizes the light of knowledge and education. Its message is to reach out, grasp the torch, share the learning, then pass it on to future generations.

CHALLENGER

BRAND GIBSON McCANDLESS McNAIR STEWART

FOR THE CREW OF STS-11
WITH ADMIRATION AND BEST WISHES

Landsat over China
False-color imaging techniques can extract a wealth of different information from data returned by remote-sensing satellites such as the U.S. Landsats. Operators can massage the data to highlight, in false color, surface features with a characteristic "spectral signature". In this Landsat image of the Luichow Peninsula in China, for example, pale blues show the presence of sediment in the water.

▶▶
A for Aeronautics
NASA scientists mastermind one of the most complex simulations ever carried out to visualize the air flow through a turbojet. This serves as a reminder that the first A in NASA stands for Aeronautics and that it was born out of the National Advisory Committee for Aeronautics, NACA.

▶▶
Psychedelic Mars
Using imaging data returned from the *Viking 2* probe, scientists at the U.S. Geological Survey at Flagstaff, Arizona, produce this vivid false-color mosaic showing variations in the surface chemistry of Mars.

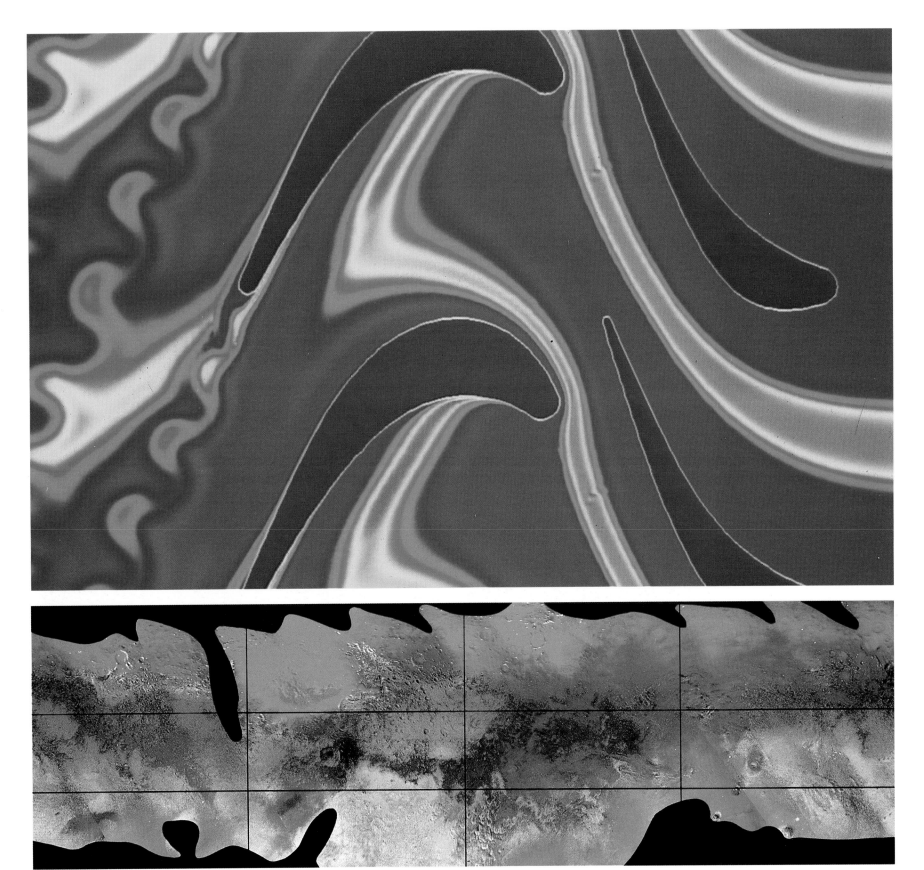

2 Establishing a foothold

"I believe that this nation should commit itself to achieving the goal, before this decade is out, of landing a man on the Moon and returning him safely to Earth."

President John F. Kennedy, speech before Congress, May 25, 1961.

The idea of traveling in space is nearly as old as civilization itself, although in ancient times the heavens were held to be the province of gods and goddesses and mythical creatures, not of mere mortals. Until the early years of this century no one really believed that human beings would be able to travel in space. In fact the necessary technology did not become available until the Cold War of the 1950s, which saw both the United States and the Soviet Union contending for deterrent supremacy by developing powerful intercontinental ballistic missiles (ICBMs).

In practice the first human beings were lofted into space in sardine-tight cans on top of ICBMs. It was not the way it should have been done, but it was the quickest. Speed was of the essence in the rivalry that existed between East and West. In the event, the Russians set the pace and the Americans had to struggle to keep up.

The Soviet launch of the first man in space, Yuri Gagarin, stung the American President, John F. Kennedy, into making his impassioned plea of 1961 for the nation to land a man on the Moon by the end of the decade. And this was before any American had even made it into orbit.

The following year one did, John Glenn, and he was feted from coast to coast as befitted the clean-living, all-American hero that he was. His was the first orbital flight of NASA's first man-in-space project, Mercury, named after the fleet-footed messenger of the gods in Roman mythology. Three other astronauts followed him into orbit over the next 15 months, the last, Gordon Cooper, remaining in space for nearly a day and a half.

Americans did not return to orbit until March, 1965, when the first manned flight of the two-man Gemini project took place. Gemini marked the turning point in what had begun to be called the Space Race. It put ten teams of astronauts into orbit over a 20-month period, one of them for nearly two weeks. During these flights NASA astronauts acquired the skills and developed the techniques they would need for the next great push into the universe — the exploration of the Moon in project Apollo.

The hardware needed to support the Apollo assaults was also being readied, as was the massive launch complex at Cape Canaveral, which evolved into the world's busiest spaceport, the Kennedy Space Center.

Astronaut no. 1
President John F. Kennedy congratulates pioneering astronaut Alan Shepard following his suborbital flight on May 5, 196l.

Astronaut no. 2
On July 21, 1961, Virgil Grissom rides a Redstone rocket as he sets off on a 15-minute suborbital flight into space and back.

Let us do something!

The Space Age was nearly a year old when NASA, the National Aeronautics and Space Administration, was founded, on October 1, 1958. It brought together all the disparate factions of the old NACA (National Advisory Committee for Aeronautics), with the common goal of developing a viable American space program, and one that would be second to none. At the time, however, the American space program was running second — to the U.S.S.R.

The Space Age had begun on October 4, 1957, with the Soviet launch of the world's first artificial satellite, *Sputnik 1*. It weighed nearly 184 pounds (84 kilograms). To prove that the launch was no fluke, the Soviets launched a second satellite, *Sputnik 2*, a month later. Not only did it weigh a colossal half a ton, it also carried the world's first space traveler, a dog named Laika. That the U.S.S.R. had the capability of launching such a heavy payload meant that their rocket technology was vastly superior to that of the United States by several orders of magnitude.

Galvanized into action by the Sputnik launches, teams from the U.S. Army and Navy vied with each other for the honor of thrusting America into the Space Age. The Army team was under the direction of the legendary Wernher von Braun, who had designed the wartime V2 rocket, ancestor of all space rockets. He had been pleading for years with the military and with politicians for an opportunity to launch a satellite, but without success. On hearing of the launch of the first *Sputnik*, von Braun stormed: "We knew they were going to do it. For God's sake, turn us loose and let us do something!"

The Navy team tried first in December 1957 with a Vanguard rocket; ignominiously, this blew up on the launch pad at Cape Canaveral. Next it was the turn of von Braun's team, with a modified Jupiter C rocket called *Juno 1*. On January 31, 1958 the rocket made a perfect launch from the Cape and put into orbit a satellite called *Explorer 1*. It weighed a mere 31 pounds (14 kilograms), just one thirty-sixth of *Sputnik 2*! Such was the gulf in propulsion technology between the two space powers.

The Original Seven

Within six days of its inception, NASA announced the commencement of a man-in-space program. It was obvious that it would only be a matter of time — probably a short time — before the Soviets attempted manned flight. Their superior rocketry practically guaranteed this. In November, 1958, the first U.S. manned project, Mercury, was initiated.

U.S. President Dwight Eisenhower decided that the

first Americans in space would be test pilots drawn from the military. Just seven men survived the exhaustive selection and screening procedures, psychological and medical testing, and were presented to the media in April, 1959. They were Gordon Cooper, Virgil "Gus" Grissom, and Donald "Deke" Slayton from the Air Force; John Glenn from the Marines; and Scott Carpenter, Walter Schirra, and Alan Shepard from the Navy. They became known as the "Original Seven" astronauts.

The Mercury capsule
The spacecraft designed to carry the first Americans into space was a cone-shaped vehicle a little under 9 feet (3 meters) tall and about 6 feet (1.8 meters) in diameter at the base. The inner pressure hull was almost as cramped as a telephone booth. Said Mercury astronaut John Glenn: "You don't climb into the Mercury spacecraft, you put it on!"

The capsule, as the cramped spacecraft was termed, was designed for fully automatic flight. It would be launched and recovered on a predetermined trajectory. The "Mercury Seven", highly trained test pilots that they were, were less than happy with this arrangement, which they called the "Spam in a can" option. And so modifications were made so that they could actually fly the craft in orbit, changing its orientation by firing the attitude thrusters.

When the time came to return to Earth, retrorockets would fire and slow down the capsule, allowing it to fall back under gravity. It would re-enter the Earth's atmosphere base first. The heat shield on the base would melt and boil away, but it would protect the capsule structure and the astronauts inside. As on all U.S. space flights before the Shuttle era, the capsule was designed for recovery at sea. The drag of the atmosphere, followed by parachute braking, would slow the craft for a gentle splashdown.

Ham in a can
Another gripe of the Mercury Seven was that they objected to monkeys and chimpanzees beating them into space! On the last day of January, 1961, a chimpanzee named Ham rode a Mercury capsule on a suborbital flight which was a dress rehearsal for a manned flight. The capsule was lofted in an arcing, ballistic trajectory that took it into space and back but without going into orbit. Ham's flight was an unqualified success.

America was ready to put a man in space, and on February 21 NASA announced its plans to the world. Perhaps it would have been better to keep quiet. Before American intentions could be put into practice the

First into orbit
John Glenn enters the cramped Mercury capsule *Friendship 7* prior to his launch into orbit on February 20, 1962. The flight takes in three orbits of the Earth and lasts nearly 5 hours.

Mercury recovery
The recovery of Mercury capsule *Sigma 7* on October 3, 1962, brings to a close the third successful U.S. orbital flight that year. The pilot is Walter Schirra.

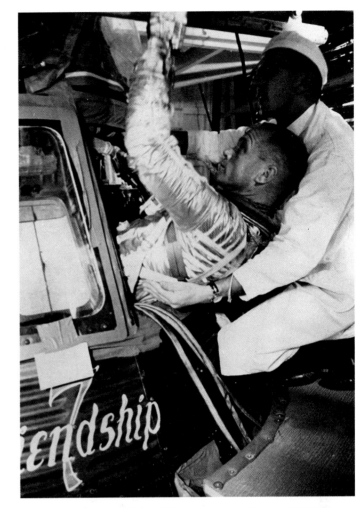

▲
Gemini lift-off
Gemini 6 starts its journey into
space on December 15, 1965.
During its one day in orbit, it
makes a rendezvous with
Gemini 7, which is on the way
to a record of nearly 14 days in
space.

Floating in space
Gemini 4 astronaut James
McDivitt photographs
colleague Edward White as he
begins a spectacular
20-minute spacewalk on
June 3, 1965.

▶
Gemini recovery
The recovery of *Gemini 6*, as
with all pre-Shuttle U.S.
manned flights, takes place at
sea. Recovery ships standing
by in the recovery area send
divers by helicopter to assist
the astronauts.

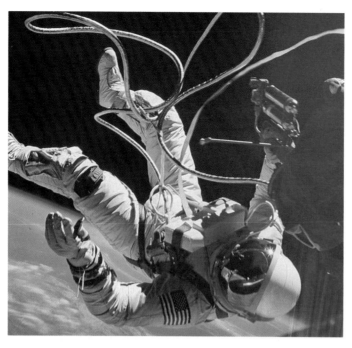

Soviets struck again by launching Yuri Gagarin into
space on April 12. This was not a suborbital flight like
the one NASA was planning, but a full-blown 90-minute
cruise in orbit.

"Space is open to us now"
It was with a sense of anticlimax that NASA launched
the first American astronaut on May 5, 1961. He was
Alan Shepard in the Mercury capsule *Freedom 7*, which
was accelerated by a Redstone rocket to 5,000 miles per
hour (8,000 km/h) and reached an altitude of 115 miles
(185 kilometers). Shepard had quite a rough ride on the
way up and endured braking forces of up to 12g on the
way down (this meant that his body would have felt 12
times heavier than usual). It was, said Shepard
laconically afterwards, "a pleasant ride".

Twenty days later President John F. Kennedy decided
that the time had come to inject a greater sense of
urgency into the American space program. In an historic
speech before Congress he urged the nation to make a
supreme effort and go for the Moon. Said Kennedy
"Space is open to us now, and our eagerness to share its
meaning is not governed by the efforts of others I
believe that this nation should commit itself to achieving
the goal, before this decade is out, of landing a man on
the Moon and returning him safely to Earth. No single
space project in this period will be more impressive to
mankind, or more important for the long-range
exploration of space."

An American in orbit
However, the Moon was a long way off in distance and
in time. First, whatever its long-term plans, NASA had
to get an astronaut into orbit. After another suborbital
flight in July, 1961, by Gus Grissom in *Liberty Bell 7*, it
was time to head for orbit. A successful two-orbit flight
by a chimpanzee named Enos in November, 1961,
prepared the ground for the first manned orbital flight
early in 1962.

After several delays because of the weather, the flight
finally got underway on February 20. At 9:47 A.M.
Eastern Daylight Time John Glenn lifted off the launch
pad in *Friendship 7* atop a modified Atlas ICBM. He
accelerated to the orbital velocity of 17,500 miles per
hour (28,000 km/h) in only five minutes. Then he was
floating free in space. Three times he sped around the
Earth before retrobraking and plunging Earthwards.
There was some concern about whether the heat shield
would withstand re-entry, but it did, and Glenn notched
up a much needed success for the American space
program.

The Mercury project called for further flights to
confirm that human beings could be launched safely into

space and survive there, at least for short periods of time. Two more flights took place in 1962. On May 24 Scott Carpenter in *Aurora 7* repeated Glenn's three-orbit mission. On October 3 Walter Schirra completed orbits in *Sigma 7*.

On May 15, 1963, Gordon Cooper blasted off in *Faith 7* on the final Mercury mission, aiming for a record (for the Americans) 22 orbits of the Earth "to determine the effects of extended space flights on man". When President Kennedy congratulated Cooper later at the White House, he was tempted to say: "I think before the end of the '60s, we will see a man on the Moon, an American." Kennedy was to be proved right. But tragically, he himself did not live to see the reality. Seven months later, on November 22, 1963, he was assassinated at Dallas, Texas. Yet his dream lived on.

In all, during the Mercury project, NASA astronauts spent two and a half days in space. But the U.S. was still lagging behind the U.S.S.R. in both space-flight experience and technology. As if to emphasize the point, a month after Cooper splashed down the Soviets stage-managed a joint flight, with two cosmonauts in orbit at the same time. One was the first woman in space, Valentina Tereshkova. Together the two cosmonauts notched up nearly eight days in orbit.

The astronauts went in two by two
When NASA first planned its manned assault on the space frontier, it intended to follow the Mercury flights with a long series of Apollo flights designed to land a man on the Moon by the mid-1970s. The Kennedy-imposed deadline of the end of the 1960s changed all this. Instead, NASA developed another project that would enable it to perfect the techniques that would be necessary for a successful Moon landing.

The project was Gemini, named after the constellation of the Zodiac, the Heavenly Twins, because it featured a two-man spacecraft. The crew compartment of Gemini was in effect a scaled-up Mercury capsule, with a similar conical shape but with room for two seats. Unlike Mercury, it had additional units attached. Immediately behind the crew capsule was an adapter section that contained equipment and consumables. At the rear was a retromodule, which housed the retrorockets.

Gemini also had a docking unit in the nose so that it could dock, or join up, with other spacecraft. Docking techniques would be required on the forthcoming Apollo missions. Gemini also had two hatches over the astronauts' seats, through which they could practice extravehicular activity (EVA), or spacewalking.

Upstaged again
The first manned fight of Gemini was finally set, months behind schedule, for March 23, 1965. As the countdown was about to begin, the Russians characteristically staged another spectacular. On March 18 cosmonaut Alexei Leonov made the world's first spacewalk from the spacecraft *Voshkod 2*.

Upstaged but not upset, Virgil Grissom and John Young blasted off as planned in *Gemini 3* on March 23. A three-orbit test flight, it went without a hitch. *Gemini 4*'s three-orbit flight in June was marked by a spectacular spacewalk by Edward White, two years later one of the first casualties of America's thrust into space.

Eight more Gemini flights took place, with all mission objectives being met. *Gemini 5* in August doubled the space duration record to nearly eight days. *Gemini 7* in December beat this by six days, and during its marathon flight rendezvoused with *Gemini 6*. In March, 1966, *Gemini 8* carried out the first docking in space, with a previously launched Agena target vehicle. This flight marked the space debut of Neil Armstrong, soon destined to take the first "small step" on another world.

The remaining Gemini crews practiced more spacewalking and more rendezvousing and docking with Agena vehicles. In September, 1966, the *Gemini 11* astronauts used an Agena rocket as a booster to lift them to the highest altitude ever, 860 miles (1,370 kilometers). For the first time it was possible to see Earth as a whole. Yelled Charles Conrad: "The world is round!" A 5½-hour spacewalk by Edwin Aldrin was the highlight of the last mission, *Gemini 12*, in November.

Ten highly successful missions, 1,940 man-hours in space, all mission objectives achieved ... the Americans had at last caught up and sprinted past the Soviets, who curiously had launched no manned spacecraft since the Gemini flights began. Were they too girding their loins for a push to the Moon?

American preparations for an assault on the Moon by the end of the decade were now forging ahead. The Apollo hardware was nearing completion, and the new launch facilities at Cape Canaveral were nearly complete. In late 1966 NASA announced that the first manned Apollo mission would be attempted in February, 1967.

Showered with honors
This work by J. M. Rosenberg honors "Colonel John Glenn and his colleagues" in March, 1962, shortly after Glenn made the first U.S. orbital flight. It shows the astronauts taking part in a triumphant motorcade.

Complex 14
The gantry at Complex 14 at Cape Canaveral, from which the Mercury astronauts were lofted into space, has long since been dismantled. In its place is a monument honoring the "Original Seven" astronauts. There is also a plaque bearing a portrait of John Glenn.

The Mercury capsule
With the escape tower attached, the bell-shaped Mercury capsule measures 26 feet (7.9 meters) long. The diameter of the heat shield at the base is about 6 feet (1.8 meters). The escape tower is designed to lift the 1-ton capsule clear of the launch rocket in an emergency.

COMPLEX - 14
MANNED ORBITAL FLIGHTS
— by —
Lt. Col. John H. Glenn Jr., USMC. 20 Feb. 1962
LCDR. M. Scott Carpenter. USN. 24 May 1962
LCDR. Walter M. Schirra Jr., USN. 3 Oct. 1962
MAJ. L. Gordon Cooper. USAF. 15 May 1963

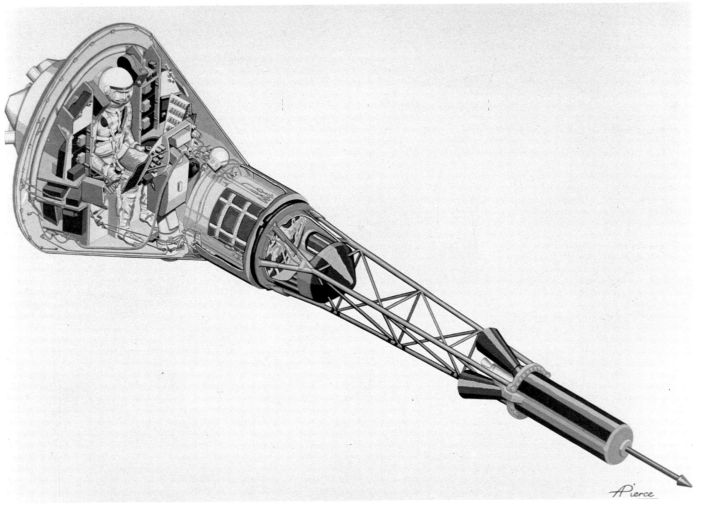

Launch complex
The beehive structure seen at
this early launch complex is the
blockhouse, from which the
launch is controlled. It is
designed to withstand blast in
the event of the launch rocket
exploding (as it sometimes did
in the early days!). Paul Sample
painted this picture from one
of the gantries.

"Support"
A view from one of the launch gantries at Cape Canaveral, painted by James Wyeth. It looks along the row of launch complexes located near the shoreline. These complexes were active during the 1960s for manned and unmanned space launches and also for the testing of ICBMs. Most have since been dismantled.

▲

"Mercury-Atlas 9"
A rescue crew, on the alert for trouble, watch the Mercury-Atlas 9 launch vehicle soar into the heavens on May 15, 1963. It carries Gordon Cooper in the Mercury capsule *Faith 7*. He is aiming to spend more than a day in space. This painting of the scene at Launch Complex 14 is the work of Robert McCall.

"After thoughts"
A spacesuited Gordon Cooper snatches a moment to reflect on his record-breaking flight, on which he saw the Sun rise and set 20 times over. The artist is Mitchell Jamieson, who sailed in the recovery ship to record the splashdown of Cooper's capsule.

"First Steps"
Gordon Cooper is pictured on
May 16, 1963, by artist
Mitchell Jamieson. He is still
in his silver spacesuit, shortly
after recovery following his
22-orbit mission in space.

"Cape from Cocoa Beach"
The sandy shores of Cocoa Beach lie a few miles south of Cape Canaveral and still, as in the early days of the space program, provide a superb vantage point for watching launches from the Cape. On a clear day the launch gantries can be seen on the skyline, as in this painting by Nicholas Solovioff.

"Shore Bird"
A seagull flies along the shore of Cape Canaveral on a stormy day, in this painting by John Willis. On such a day as this only the birds will be flying. The rockets will remain on their launch pads.

"Wet Morning"
The Cape enjoys a hot and humid subtropical climate, and is covered by large swampy areas where wildlife abounds. This ambience is well captured in this painting by Hugh Laidman, which shows one of the antennas used for tracking launch vehicles.

▶

Ready Room 1
In June, 1965, the U.S.S. *Wasp* is assigned as the recovery vessel to pick up *Gemini 4* astronauts Edward White and James McDivitt after splashdown. Here in Ready Room 1 the crew receive a final briefing on recovery operations. Artist Franklin McMahon is on hand to record the scene and also the recovery.

***Gemini 4* recovery**
Minutes after the *Gemini 4*
capsule splashes down on June
7, 1965, divers fly in from
U.S.S. *Wasp* with a flotation
collar to attach to the capsule
and a dinghy to take the crew.
Here, divers prepare to help
astronauts Edward White and
James McDivitt open the
hatches.

Red carpet treatment
The *Gemini 4* astronauts get
the red carpet treatment
aboard the U.S.S. *Wasp.* Their
62-orbit flight marks a
milestone in U.S. space flight,
for it sees the first spacewalk by
an American, Edward White,
destined to become one of
America's first space martyrs.

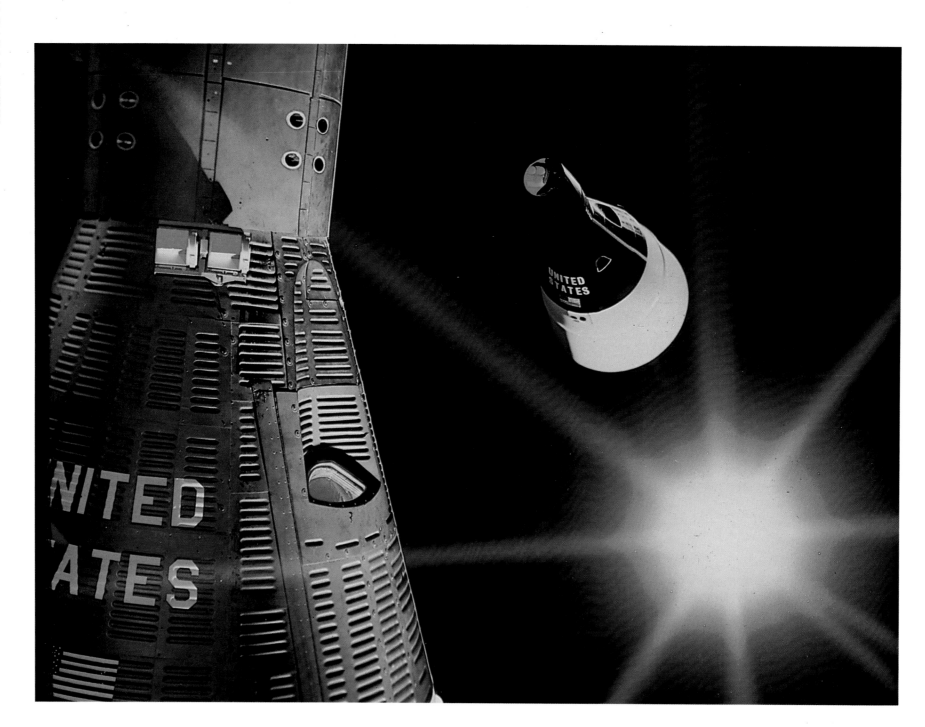

◄ Rendezvous in orbit

Rendezvous maneuvers between spacecraft are essential to the success of the Apollo Moon-landing missions. Two Gemini spacecraft, *6* and *7*, carried out the first in-orbit rendezvous on December 15, 1965, previewed in this pre-mission photograph, which uses a full-scale and a one-sixth scale spacecraft.

Saturn hardware

While the Gemini flights are perfecting the techniques that will be required for the Apollo missions, the necessary hardware is being readied. Wernher von Braun's team spearheads launch-vehicle design at the Marshall Space Flight Center in Alabama, where artist Howard Koslow painted this scene at the time of the first Gemini mission in March, 1965.

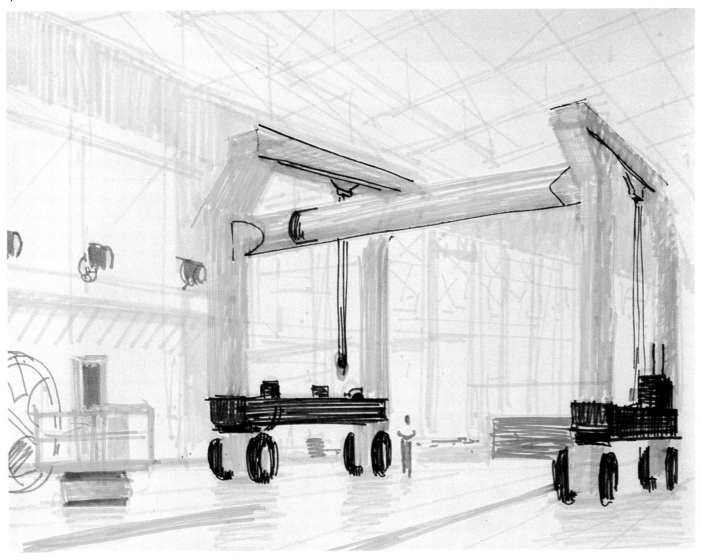

"From the House"
The Gemini missions, launched by Titan rockets, lift off at Launch Complex 19 at the Cape. A few hundred yards north are Complexes 34 and 37, where the first and second generation Saturn rockets, I and IB, are launched. Robert Shore painted this picture of the Saturn gantry, showing the massive girder work, coated with red corrosion-resistant paint.

Rising from the swamp
Alfred McAdams' painting shows an early stage in the construction of the Vehicle Assembly Building (VAB). The VAB begins to rise above the flat Florida landscape in 1963 and is completed three years later. Towering 526 feet (160 meters) high, it stands as a monument to the success and scale of the U.S. space program.

3
Destination Moon

"The Moon is within reach. The trip will not be made tomorrow, but many rocket experts believe it will not be long."

Excerpt from "A trip to the Moon and back", Sacramento Bee, *February 13, 1947.*

While Gemini astronauts were notching up impressive records in orbit in 1965 and 1966, other spacecraft were being dispatched to the Moon to reconnoiter suitable landing sites. First came the Ranger probes, which sent back close-up TV pictures of the lunar surface before crash-landing. Then came the Surveyors and Lunar Orbiters. *Surveyor 1* soft-landed on the Moon on June 1, 1966, sending back the most detailed pictures yet and proving that the surface was firm enough to land on. The Lunar Orbiter probes, beginning in August, went into orbit around the Moon and mapped it in detail. From the pictures they sent back, the sites for the Apollo landings were selected.

The rocket that would carry men to the Moon, the Saturn V, was meanwhile nearing completion. It stood 36 stories high on the launch pad. Nothing like it had ever flown before, and after the Moon shots nothing like it would ever fly again. The gigantic rocket was designed to launch a crew of three to the Moon in a three-module spacecraft. Only two of the three astronauts would land and explore. To house the monstrous Saturn V, a suitably gargantuan building was being completed at Cape Canaveral, the 50-story-high Vehicle Assembly Building (VAB).

The first test flight of the Apollo spacecraft was scheduled for February, 1967, but in the event it did not take place until 20 months later, in October, 1968. It was delayed by the need to redesign the crew module of the Apollo spacecraft following a tragic fire in January, 1967, which killed three astronauts in training.

But from the time of the first test flight, designated *Apollo 7*, everything went swimmingly with the Moon-landing build-up. *Apollo 8* made a circumnavigation of the Moon. *Apollo 9* and *10* rehearsed the Moon-landing procedures with complete success. And on July 16, 1969, *Apollo 11* set off on its historic journey. Four days later Neil Armstrong took the first steps on the Moon.

Five more teams of Apollo astronauts made lunar landings. In total they explored the Moon in EVAs totalling more than 80 hours' duration, roaming over dusty plains and rugged highlands. They brought back with them 850 pounds (385 kilograms) of Moon rocks and soil, and a portfolio of photographs recording the greatest adventure there has ever been — yet.

The Saturn family

Wernher von Braun's Army-derived team of rocket experts began designing heavy launch vehicles for Earth orbit even before the Apollo project was underway. They were based at the Marshall Space Flight Center at Huntsville, Alabama. The result was the Saturn series of launch vehicles. The one that would launch men to the Moon was the Saturn V, developed using technologies and components perfected on earlier vehicles, particularly the Saturn IB.

The Saturn IB was a two-stage rocket with a distinctive cluster of propellant tanks feeding kerosene and liquid oxygen to the eight engines of the first stage (the S-IB). The upper stage (the S-IVB) had a single engine, which burned liquid hydrogen and liquid oxygen. This was a novel combination of propellants at the time. The first flight of the Saturn IB lifted off on February 26, 1966. Further flights took place to test-fly components for the Apollo spacecraft and to launch the first manned Apollo mission (*Apollo 7*) into orbit.

The only thing the Saturn IB and the Saturn V had in common was the upper stage, the S-IVB. The Saturn V was unbelievably huge. Its first stage alone, 150 feet (46 meters) long, was bigger than the complete Saturn IB. Overall, with the Apollo spacecraft attached, the Saturn V stood 365 feet (111 meters) high on the launch pad. With a full load of propellants the vehicle weighed the best part of 2,900 tons!

The first wholly successful flight of Saturn V took place on November 9, 1967, from the newly built launch pad at Complex 39 at Cape Canaveral. A second test flight came in April, 1968. The next time it flew, it was to the Moon, with *Apollo 8*.

Rendezvousing in lunar orbit

The reason such a huge launch vehicle was needed for the Moon shots was that the Apollo spacecraft weighed 45 tons. This was a huge weight to heft into orbit, let alone propel at an escape velocity of 7 miles (11 kilometers) per second.

The Apollo spacecraft had a modular design that was dictated by the method NASA chose to reach the Moon. This method was lunar orbit rendezvous. The Saturn V would launch the spacecraft into lunar orbit. A landing vehicle would then separate and take the astronauts down to the Moon's surface. On return from the surface, the astronauts would rendezvous and dock with the mother ship in lunar orbit, and return home in that.

The Apollo spacecraft

Three modules made up the Apollo spacecraft. A crew of three astronauts occupied the command module, a cone-shaped unit 12 feet (3.5 meters) high and 13 feet (3.9 meters) in diameter at the base. It was pressurized with oxygen at about one-third sea-level atmospheric pressure. The base was covered with a thick reinforced plastic heat shield, which would prevent the craft and the crew from burning up during re-entry.

For most of the journey to and from the Moon the command module was mated with the service module, a unit that housed the spacecraft's main engine and propellant tanks. It also housed the fuel cells that provided power and also drinking water for the crew. The service module carried four sets of attitude-control thrusters to permit maneuvering in space. The command and service modules were collectively referred to as the CSM. Together they contained two million components.

The third section of the Apollo spacecraft was the lunar module. This was the craft designed to separate and land on the Moon. It was an odd-looking contraption of severely functional shape, made up of two parts, an ascent stage and a descent stage. Each was fitted with an engine. The astronauts were to return to orbit in the ascent stage, using the descent stage as a launch pad.

Assembly and rollout

The Apollo spacecraft and Saturn V launch vehicle were put together in the mammoth Vehicle Assembly Building, which was the focus of the launch site for Apollo, Complex 39. The VAB was completed in 1966 after four years' work. Its dimensions were huge — 526 feet (160 meters) high, 518 feet (158 meters) wide, and 716 feet (218 meters) long. With these dimensions it could accommodate four Saturn Vs at once, one in each of its four bays. Technicians worked on the rockets from platforms located at various levels.

The 365-foot (111-meter) Apollo/Saturn V stack was assembled on top of a mobile launcher, which carried a gantry that gave access to the stack on the launch pad. The total weight of the completed stack and the launcher was around 5,500 tons! And this somehow had to be transported 3.5 miles (5.5 kilometers) to the launch pad. No existing vehicle could handle such a load, and so two eight-track crawler transporters were built for the purpose. They were the largest land vehicles ever, measuring 130 feet (40 meters) long and nearly 115 feet (35 meters) wide. They carried the Apollo/Saturn V stack to the launch pad at 1 mile per hour (1.6 km/h).

Most of the facilities at Complex 39 have survived into the Shuttle era. The VAB, the mobile launchers, the crawler transporters, the two launch pads — all have been modified and expanded as necessary to handle the present generation of Shuttle hardware.

Wernher von Braun
In the early 1940s, Wernher von Braun led the team that developed the German V2 rocket at Peenemünde in the Baltic. He began working for the Americans in 1946, immediately after World War II, and in the early 1960s became the chief architect of the Saturn launch vehicles designed to carry astronauts to the Moon.

Apollo 1 **crew**
The prime crew for *Apollo 1*, which was to have been the first manned Apollo flight: pioneer spacewalker Edward White (left); veteran of Mercury and Gemini missions, Virgil Grissom (middle); and rookie Roger Chaffee. They are destined never to ride together into space.

◄
Saturn at sunset
Sitting on the launch pad at Complex 39 and ready for its maiden flight is the Saturn V rocket. The 365-foot (111-meter) high colossus takes off on November 9, 1967, carrying an unmanned Apollo spacecraft, which is successfully recovered from orbit.

"Fire, I smell fire"

The launch date for the first manned Apollo flight, designated *Apollo 1*, was set for February 21, 1967. On the afternoon of January 27, the three astronauts who had been selected for the mission took part in a simulated countdown on the launch pad of Complex 34 at Cape Canaveral. They were Gus Grissom, a veteran from the Mercury and Gemini projects; pioneering spacewalker Edward White; and Roger Chaffee, who had yet to fly in space.

They were strapped in the *Apollo 1* command module atop the rocket that would blast them into the heavens in three weeks time. They were suited up as for lift-off and breathing pure oxygen, which also filled the capsule. Things did not go smoothly that afternoon. There were problems with communications, which prompted a frustrated Grissom to remark: "How the hell can we get to the Moon if we can't even talk between two buildings?" There were also problems with the oxygen system. But there was worse to come — *the worst*.

At half past six Chaffee radioed to launch control: "Fire, I smell fire." A few seconds later came a panic-stricken cry: "We've got a fire in the cockpit. We're burning up here." There was frantic movement inside the capsule, then it burst open in a cloud of flame and smoke. By the time pad technicians reached the astronauts, they had suffocated to death. Gus Grissom, Edward White, and Roger Chaffee had become the first of America's space martyrs. They would not be the last.

Getting back on course

The board of inquiry set up to investigate the tragedy identified an electric spark as the most likely cause of the conflagration. It set fire to insulation and from then on disaster was inevitable. With the pure oxygen atmosphere to feed on, the fire flashed through the capsule with lightning speed, consuming everything combustible.

The board recommended more than 1,300 changes in the design of the capsule and operating procedures. It was not until the fall of 1968, some 20 months later, that a newly designed command module was ready.

On October 11, *Apollo 7* astronauts Walter Schirra, Donn Eisele, and Walter Cunningham rode a Saturn IB rocket into orbit. Over the next 11 days they tested the Apollo command service modules (CSM) with great success, and even practiced simulated docking maneuvers with a discarded rocket stage. The crew also became TV stars by making the first broadcasts from space. The Earthbound were given a taste of what life was like in orbit.

Moon, Earth, and *Eagle*
The lunar module *Eagle*, containing *Apollo 11* astronauts Neil Armstrong and Edwin Aldrin, is snapped by CSM pilot Michael Collins as the two craft rendezvous in lunar orbit after the first successful Moon landing on July 20, 1969. Soon the craft will dock, and *Eagle*'s two astronauts will transfer to the CSM for an uneventful journey home.

In quarantine
The *Apollo 11* astronauts are placed in quarantine after their return from the Moon. This is a precaution to prevent the spread of any killer germs the Moon might conceivably harbor. Caged, but happy, in their quarantine module are (from left to right): Edwin Aldrin, Neil Armstrong, and Michael Collins.

Pinpoint landing
In November, 1969, *Apollo 12* astronaut Charles Conrad pilots the lunar module to a perfect touchdown within a few hundred feet of probe *Surveyor 3,* which landed on the Moon in 1967. The astronauts cut off pieces from the probe to take back to Earth for analysis.

Escaping Earth's clutches

So well had the first Apollo flight gone that NASA decided to pull out all the stops and send the next one on a pathfinding mission to the Moon. The astronauts would not land, just circumnavigate the Moon and return. This would enable the Apollo navigation team to fine-tune the celestial mechanics for an accurate landing.

Accordingly, on December 21, 1968, one year and ten days before Kennedy's deadline was up, *Apollo 8* soared from Launch Pad A at Complex 39 atop a Saturn V launch vehicle. Aboard were Frank Borman, James Lovell, and William Anders, bound where no man had gone before, at a speed no man had yet flown — nearly 25,000 miles per hour (40,000 km/h).

Three days and 240,000 miles (385,000 kilometers) later, on Christmas Eve, they were in lunar orbit. Traveling about 70 miles (110 kilometers) above the Moon, they transmitted live TV pictures of the barren, crater-scarred, but awesomely beautiful surface of our nearest cosmic neighbor. It made compulsive viewing. On one telecast the crew read movingly from the Book of Genesis: "In the beginning God created the Heaven and the Earth.... And God saw that it was good."

On Christmas morning they carried out a successful burn of the service module's engine to put them on course for home. On December 27 they hurtled into the Earth's atmosphere at 35 times the speed of sound. The heat shield of their command module did its job well, and the crew survived to make a gentle splashdown in the Atlantic Ocean. The human conquest of the Moon had begun.

Test-flying the lunar module

The Saturn V and the Apollo CSM had now proved themselves in space. As yet, the remaining piece of Apollo hardware, the lunar module, had not. This was remedied on the *Apollo 9* flight into Earth orbit on March 3, 1969. The crew put the lunar module through its paces, practicing the rendezvous and docking maneuvers that would be necessary on a flight to the Moon.

CSM pilot David Scott docked with the lunar module and drew it clear of the third-stage rocket. Later, James McDivitt and Russell Schweickart entered the lunar module and separated from the CSM. Since the two modules were now independent spacecraft, they were accorded different callsigns, *Gumdrop* for the CSM and *Spider* for the lunar module. *Spider*'s systems worked perfectly and the re-docking with the CSM went off without a hitch.

Apollo 9's successful ten-day mission paved the way for a final dress rehearsal, by *Apollo 10*, for a Moon landing. *Apollo 10* set off on May 18, 1969, to repeat in lunar orbit what had been accomplished in Earth orbit by *Apollo 9*. Thomas Stafford and Eugene Cernan in the lunar module (callsign *Snoopy*) separated from John Young in the CSM (*Charlie Brown*) and swooped down to within 50,000 feet (16,000 meters) of the lunar surface. So near and yet so far. A successful ascent-stage burn and docking put the three astronauts back together again for an uneventful trip home.

"The *Eagle* has landed"

By the time the *Apollo 10* command module splashed down, on May 26, 1969, the hardware for the first Moon-landing attempt by *Apollo 11* was already in place on the launch pad. The portents were good. A practice countdown on July 3 threw up no problems, and NASA decided to "go for launch" at the next launch window on July 16.

At 9:32 A.M. EDT on that day the five first-stage engines of the Saturn V/*Apollo 11* launch vehicle roared into life and lifted the huge rocket off the pad. Strapped in their seats in the command module 330 feet (100 meters) above the ground were the dramatis personae of this immaculately staged-managed spectacular. They were commander Neil Armstrong, CSM pilot Michael Collins, and lunar module pilot Edwin "Buzz" Aldrin.

Twelve minutes later *Apollo 11*, with the third-stage rocket still attached, was in orbit. Twenty-two minutes after noon, the third-stage engine fired again and blasted the spacecraft towards the Moon. By the afternoon of July 19, *Apollo 11* was in lunar orbit, between 62 and 75 miles (100 and 120 kilometers) above its cratered surface.

The next day, Armstrong and Aldrin in the lunar module (callsign *Eagle*) separated from Collins in the CSM (callsign *Columbia*) and dropped down towards the Moon. Taking over manual control just before landing, Armstrong maneuvered the lunar module towards a clear spot on the dusty, rock-strewn lava plain known as the Sea of Tranquillity.

With only half a minute's fuel remaining, the lander touched down. Radioed Armstrong: "Houston, Tranquillity Base here. The *Eagle* has landed." "Roger, Tranquillity," acknowledged Mission Control, Houston. "We copy you on the ground. You've got a bunch of guys about to turn blue. We're breathing again. Thanks a lot!"

Lunar touchdown occurred at 4.18 P.M. EDT on July 20, 1969. NASA had beaten President Kennedy's seemingly impossible deadline, with five months of the decade remaining. Some six hours after touchdown Armstrong emerged from the lunar module. An estimated one-fifth of the world's population, watching transfixed on their TVs back on planet Earth, saw him plant the first human footprint on an alien world. "That's one small step for a man, one giant leap for mankind," he said.

Aldrin soon joined Armstrong, and the two began their well-practiced tasks of taking rock samples, setting up scientific equipment, and breaking out the Stars and Stripes. With the stiffened flag "flying" over the Moon, they received "the most historic telephone call ever made" from President Richard Nixon at the White House. Said the President: "For one priceless moment in the history of man, all the people on this Earth are truly one."

After moonwalking for some 2½ hours, the astronauts returned to the lunar module for a well-deserved rest. In the early afternoon of July 21 they blasted off the Moon using the descent stage of *Eagle* as a launch pad. It is still there on the Sea of Tranquillity, a monument to a magnificent feat of exploration. Attached to one of the legs is a plaque that reads: "Here men from planet Earth first set foot on the Moon. July 1969 AD. We came in peace for all mankind."

"Guess what I see?"

Apollo astronauts went in peace for all mankind a second time before the decade was out. *Apollo 12* lifted off on November 14, aiming for another lunar plain called the Ocean of Storms. The launch could hardly have been more dramatic — the mammoth rocket was struck by a lightning bolt as it ascended into thick cloud. Some of the instruments aboard *Apollo 12* were affected, but there was nothing malfunctioning seriously enough to abort the mission.

Charles Conrad flew the lunar module (callsign *Intrepid*) to a perfect pinpoint landing, within 600 feet (180 meters) of an earlier visitor to the Ocean of Storms, the probe *Surveyor 3*. Even Conrad was staggered, exclaiming to Mission Control: "You'll never believe it. Guess what I see sitting on the side of the crater?"

As well as rock sampling, Conrad and fellow moonwalker Alan Bean set up the first scientific station on the Moon. This was made up of a collection of instruments called ALSEP, the Apollo lunar surface experiments package. Like the other ALSEP stations set up on subsequent missions, it continued to radio back data to Earth for years.

"We've got a problem"

The 11th day of April, 1970, saw *Apollo 13* blast off Cape Canaveral, bound for a highland region of the Moon near Fra Mauro crater. Aboard were James Lovell, Fred Haise, and John Swigert. On April 13, just after a live telecast to Earth, the crew heard "a pretty large bang". Reported Swigert to Mission Control: "Hey, we've got a problem here!" They certainly had. An explosion had ripped through the service module of the CSM (callsign *Odyssey*), robbing the spacecraft of power and oxygen.

Apollo 13 was by now over 200,000 miles (320,000 kilometers) from Earth and in a life-threatening situation. Only by using the lunar module (callsign *Aquarius*) as a "lifeboat" were the crew able to survive — just. NASA recorded the mission as a failure, but "the most successful failure in the annals of space flight."

Wheels on the Moon

A redesign of certain systems to prevent a recurrence of the problem that crippled *Apollo 13* delayed the *Apollo 14* mission until January 31, 1971. It too was aiming for a touchdown at Fra Mauro and this time succeeded. While Stuart Roosa remained in orbit in the CSM (callsign *Kitty Hawk*), Alan Shepard and Edgar Mitchell landed in the lunar module (*Antares*). During their 9½ hours of EVA the moonwalkers set up another ALSEP station and roamed around more freely than their predecessors. This was made possible by their "golf cart", a wheeled contraption officially known as a modularized equipment transporter.

The astronauts had proper "wheels" on the next mission, *Apollo 15*, which began on July 26, 1971. They took the form of a battery-powered, four-wheeled buggy called the lunar roving vehicle or rover. The lunar rover carried the moonwalkers, David Scott and James Irwin, some 17½ miles (28 kilometers) around the most dramatic lunar landscape yet. The landing site was located in the rolling foothills of the Apennine Mountains, a range with peaks soaring to 15,000 feet (4,500 meters) at the edge of the Sea of Showers.

The seventeenth-century Italian scientist Galileo would have been delighted by an experiment Scott performed. He dropped a geological hammer and a feather from the same height. They both hit the ground at the same time, exactly as Galileo's law of falling bodies predicted. On Earth air resistance affects such an experiment.

The end of an era

The two final flights in the historic Apollo series took place in 1972 amid equally dramatic landscapes. After lift-off on April 16, *Apollo 16* set down in the lunar highlands about 150 miles (250 kilometers) southwest of the *Apollo 11* landing site, near a crater called Descartes. *Apollo 17* landed in a valley close to the Taurus Mountains on the eastern edge of the Sea of Serenity. Mission scientists selected this location because there was evidence of fresh lava flows there and also debris broken off the high cliffs that boxed in the landing site.

Apollo 17 was the longest and scientifically the most rewarding of all the missions, yielding among other things 250 pounds (115 kilograms) of rock and soil samples. To a large degree the excellence of the data was due to the presence of Harrison "Jack" Schmitt, a trained geologist, for whom a visit to the Moon was the field trip of the century. He was the forerunner of a new breed of scientist-astronauts. Today scientists of many kinds venture into space aboard the Shuttle as payload specialists.

With mission commander Eugene Cernan, Schmitt roamed the ethereally beautiful valley for a total of 22 hours, traveling 22 miles (32 kilometers) in the lunar rover. Cernan was the last human being to set foot on the Moon. As he stepped up the ladder into the *Apollo 17* lunar module on December 14, 1972, he gave a promise: "We leave as we came and, God willing, we shall return with peace and hope for all mankind."

On December 19 Cernan, Schmitt, and CSM pilot Ronald Evans splashed down in the Pacific Ocean. Project Apollo, the greatest adventure in the history of mankind, conceived and executed at a cost of $25 billion, was over.

Sons of Apollo

As originally planned, there should have been another three lunar landing missions, *Apollo 18*, *19* and *20*. But budget cutbacks and diminishing public interest in space projects put paid to them. NASA thus had surplus Apollo hardware available, which enabled it to launch two spin-off projects, *Skylab* and ASTP. *Skylab* was a hugely successful experimental space station, visited by three teams of astronauts in 1973/74 (see page 126).

The ASTP was the Apollo Soyuz Test Project, a cooperative mission with the U.S.S.R. in 1975. It was born as a result of an agreement covering "cooperation in the exploration and use of space for peaceful purposes" signed by U.S. President Richard Nixon and Soviet Premier Alexei Kosygin in May, 1972.

The crews selected for this unique mission were astronauts Thomas Stafford, Vance Brand, and Donald Slayton, and cosmonauts Alexei Leonov, the pioneer spacewalker, and Valery Kubasov. The two teams of astronauts and cosmonauts would fly into orbit in their respective spacecraft — an Apollo CSM and a Soyuz —

Ruptured tanks
Graphic evidence of the explosion that crippled the *Apollo 13* spacecraft can be seen in this photograph of the service module, taken as it was jettisoned just prior to re-entry.

and link up for nearly two days to exchange crews and conduct experiments together.

The mission began on July 15, 1975. Soyuz lifted off first, from the Baikonur Cosmodrome in Central Asia. Apollo blasted off from Cape Canaveral seven hours later. It carried a module fitted with a docking port which was compatible with Soyuz's docking mechanism. Two days later the two craft edged together and finally docked. Said Apollo commander Stafford in Russian: "We have succeeded. Everything is excellent." Replied Leonov in English: "Well done, Tom. It was a good show." Soon the two crews were meeting face to face, shaking hands in a gesture symbolizing respect and friendship.

The joint venture was a complete success. Only one thing marred the American mission. At splashdown on July 24 the crew were exposed to danger from the toxic gas nitrogen tetroxide seeping out from the command module's attitude-control thrusters. Brand lost consciousness for a minute but like the rest of the crew suffered no permanent ill-effects.

So the Apollo era finally drew to a close. The Apollo-Soyuz flight also marked the end, for the U.S. at any rate, of the "expendable era" of manned space flight. When American astronauts next flew into orbit, it would be in a revolutionary new aerospace machine that could be used again and again — the Space Shuttle. But that would not be for another six years.

The Moon tomorrow

Cernan's promise that, God willing, man would return to the Moon, should be fulfilled some time after the turn of the next century. Many schemes and likely scenarios for the setting up of a moonbase, scientific outposts, and mining facilities on the Moon have been advanced. Nothing concrete is likely to be established in the present decade, however, while NASA struggles to get space station *Freedom* operational. When this happens, time and resources will be more readily available to support a new lunar initiative. And when astronauts do return to the Moon, it will not be for a fleeting visit. This time it will be for good.

Welcome return
After four agonizing days when they were all but marooned in space, the crew of *Apollo 13* are welcomed aboard the recovery ship U.S.S. *Iwo Jima* just after splashdown in April, 1970. They are (from left to right): Fred Haise, John Swigert, and James Lovell.

Moonwalker and buggy
This powerful photograph shows *Apollo 17* astronaut Eugene Cernan with the lunar rover on the final Apollo mission, in December, 1972. Powered by batteries, the rover had a top speed of 10 miles per hour (16 km/h) and was developed at a cost of $40 million.

▲
Apollo-Soyuz
Pictured on the historic joint flight of the Apollo and Soyuz spacecraft in July, 1975, are Soviet crew commander Alexei Leonov (left) and U.S. docking module pilot Donald Slayton. Who is upside-down?

Launch Complex 37
This launch complex is the farthest north of the row of complexes that runs parallel to the shoreline at Cape Canaveral. It was built to launch the Saturn IB rocket, a test vehicle for Apollo/Saturn V hardware. Theodore Hancock's painting of the complex includes, at right, the beehive-like blockhouse housing the control center.

"Power"
The first-stage engines of the Saturn V launch vehicle roar into life and begin to lift nearly 3,000 tons of metal and propellants off the launch pad.

Paul Calle's dramatic painting captures the raw power the engines unleash as they burn kerosene and liquid oxygen at the rate of 15 tons per second.

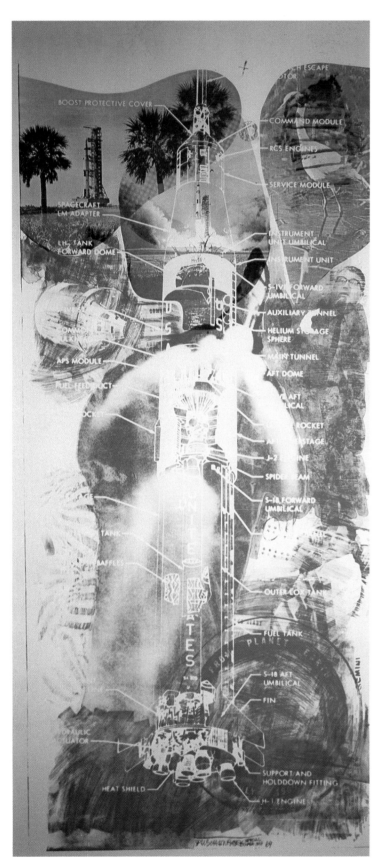

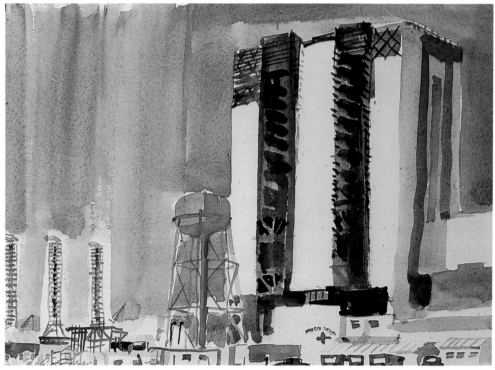

"Sky Garden"

This celebrated work by Robert Rauschenberg takes its inspiration from the gargantuan Apollo/Saturn V launch vehicle. The vehicle is superimposed as a ghosted cutaway on a variety of related scenes, for example, views of operations at the Kennedy Space Center, such as rollout and lift-off. Note at top right the portrait of Saturn's designer, Wernher von Braun.

The giant hanger

In 1965 the mammoth Vehicle Assembly Building (VAB) on Merritt Island is nearly complete. Paul Arlt's painting of the building shows, at left, three of the launch umbilical towers to which the Saturn V rockets will be mated prior to launch. The scale of the VAB can be gauged from the size of the car in the foreground.

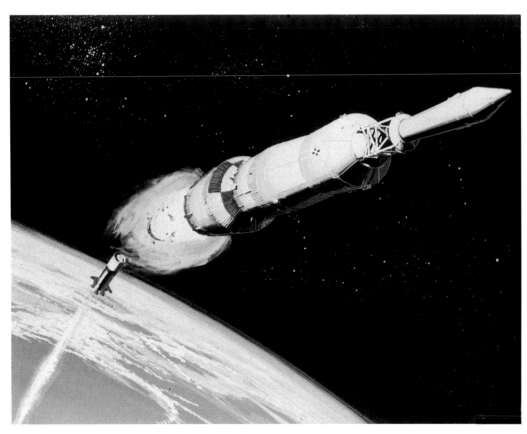

First-stage separation
The pictures on the next pages show essential stages in the flight of the Apollo spacecraft from lift-off to splashdown. The first-stage engines of the Apollo/Saturn V launch vehicle burn for just 2½ minutes and produce a thrust of 7.5 million pounds (3 million kilograms). Then they run out of fuel, and the first stage drops away. The second stage now fires and will in turn drop away, allowing the third stage and the attached Apollo spacecraft to climb into orbit.

Lunar insertion
The Apollo spacecraft, with the third stage still attached, goes into a parking orbit around the Earth, while all systems are checked out and navigational details are finalized. Then the third stage fires again to boost the speed to escape velocity, 25,000 miles per hour (40,000 km/h).

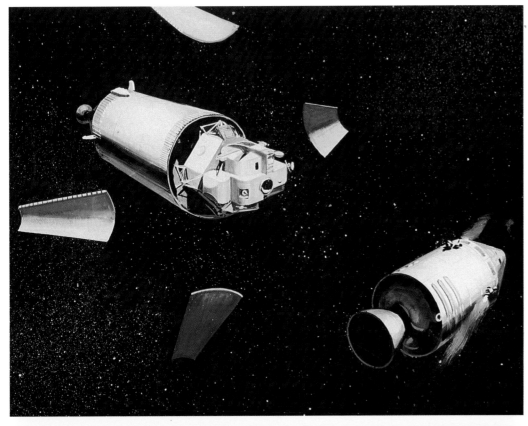

About turn
Half an hour after the second third-stage burn, the astronauts start maneuvers to configure the Apollo spacecraft for the lunar cruise and landing phases. The CSM separates, and the adapter panels are jettisoned, exposing the lunar module (LM) inside the third stage. The CSM fires its thrusters to perform a 180-degree turn.

Lunar module docking
The CSM then moves gently in to dock with the LM. It then fires its thrusters and pulls away from the third stage. The Apollo spacecraft is now in its correct configuration for lunar landing.

Retrobraking
About two and a half days after leaving Earth, the Apollo spacecraft nears the Moon. As it swings around the far side, the astronauts fire the service module's engine in the forward direction so that it acts as a brake. This maneuver insures that the spacecraft will slow down and go into lunar orbit.

Lunar touchdown
For the lunar landing phase, two astronauts transfer to the LM. They separate from the CSM and use their descent engine as a retrobrake to slow them down below lunar orbital velocity. They use retrobraking again in order to make a soft landing on the lunar surface. After exploring, they take off in the upper stage of the LM, using the lower one as a launch pad.

Red-hot re-entry
The Apollo astronauts return home in the CSM, having discarded the LM in lunar orbit. Just before they re-enter Earth's atmosphere, they jettison the service module and plunge into the upper air in the command module (CM) at a speed of 7 miles per second (11 km/s). The heat shield blazes red-hot as drag slows them down.

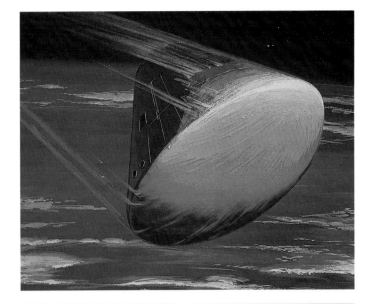

Parachute splashdown
As the CM falls lower, first small drogue parachutes and then huge main parachutes open to lower the CM to a gentle splashdown at sea.

***Apollo 10* recovery**
On May 26, 1969, the *Apollo
10* command module is
hoisted aboard U.S.S.
Princeton after a successful
dress rehearsal for the
upcoming lunar landing
mission by *Apollo 11*. Artist
Tom O'Hara is present to
record the scene.

"Go *Apollo 11!*"
It is a little after 9:30 A.M. local time at Cape Canaveral on July 16, 1969, and *Apollo 11* has begun its historic journey to the Moon, in this painting by John Meigs. The spacecraft is carrying Neil Armstrong, Edwin Aldrin, and Michael Collins on the greatest adventure of all time. In four days there will be human footprints on the lunar soil.

Man on the Moon
This painting by Norman Rockwell depicts an astronaut's first step in lunar soil. On July 20, 1969, Neil Armstrong stepped onto the Moon, watched by an estimated 500 million people back on Earth. He was the first of a dozen astronauts to do so over the next two and a half years.

▼
Lunar reflections
The visor of the Apollo spacesuit is gold-tinted to shield the astronaut's eyes from the dazzling glare of the Sun. It makes a perfect mirror, which prompts Neil Armstrong to take the most famous of all space photographs. It shows Edwin Aldrin posing in the desolate lunar landscape with the photographer and the *Apollo 11* lunar module reflected in his visor.

"Lunar Confrontation"
Artist Robert Shore exploits the same idea in this work, which sees the 19th-century French science-fiction writer Jules Verne reflected in the golden visor. Verne journeyed often to the Moon in his imagination.

Mission Control
Flight controllers at Mission Control, Houston, are riveted to their consoles on July 20, 1969, as they monitor the descent of *Apollo 11* to the lunar surface, in this drawing by Franklin McMahon. At 3:17 A.M. Houston time Armstrong reports: "Tranquillity Base here. The *Eagle* has landed."

"Hey, Houston, we've got a problem."
The call to Mission Control by the *Apollo 13* crew on April 13, 1970, prefaces the first in-flight emergency in the U.S. manned space program. An explosion has knocked out the crew module, forcing the astronauts to use the lunar module to survive. Here, artist Joseph C. Chizanskos pictures the grim-faced crew passing through the access tunnel between the two modules.

Flying the flag
An astronaut plants the U.S. flag in the lunar soil. It is specially stiffened to "fly" in the airless environment. In the astronaut's other hand is a tool for collecting rock samples. The figure is a detail in a 75-foot (23-meter) long mural at the National Aeronautics and Space Museum in Washington D.C. Entitled "The Space Mural — A Cosmic View", it was painted by Robert McCall.

Lunar roving
Another part of the mural depicts a second astronaut in front of the lunar roving vehicle, or Moon buggy. At right is the lunar module, wrapped in gold foil in places as protection against solar radiation.

Lunar lift-off
The astronauts lift off the surface of the Moon in the upper stage of the lunar module. From *Apollo 15* on, the moment of lift-off is captured in a kaleidoscope of color by the TV camera mounted on the lunar rover.

◀

Historic handshake
Up in orbit on July 17, 1975, U.S. astronauts and Soviet cosmonauts shake hands during the unique ASTP (Apollo-Soyuz Test Project). At left is the Apollo CSM, which carried the docking module into orbit. This module is fitted with a port compatible with Soyuz's docking mechanism.

ASTP recovery
After a five-day mission, the American ASTP crew of Thomas Stafford, Vance Brand, and Donald Slayton splash down in the Pacific on July 24, 1975. Artist Roger Arno painted this scene from sketches he made aboard the recovery vessel U.S.S. *New Orleans*.

Lunar ferry

A lunar ferry prepares to depart for the Moon early during the next century, when a Moon base is under construction. On its return the ferry will use aerobraking in the outer fringes of Earth's atmosphere to reduce its speed so that it can re-enter Earth orbit. Other hardware featured in this scene include a space station and free-floating empty Shuttle propellant tanks, which are used for in-orbit storage.

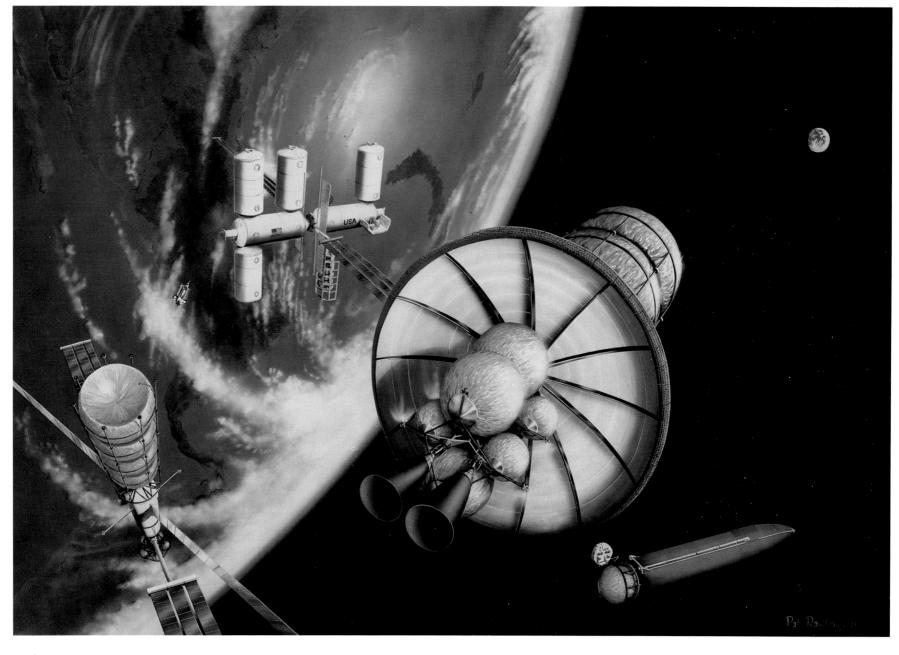

▶
Lunar supply base
Eventually the Moon will provide raw materials for space manufacturing facilities in orbit. The mined materials will be transported into space by an ingenious electromagnetic "cannon" known as a mass driver. The picture shows this "cannon", together with habitation and workshop modules for the operating personnel.

Lunar landing facility
In this artist's concept, a lander has just set down on a gravel-covered landing pad close to the Moon base. While technicians begin to drain the propellant tanks and check out the vehicle, the crew transfer through a tunnel to a pressurized vehicle (right foreground) for transport to the base.

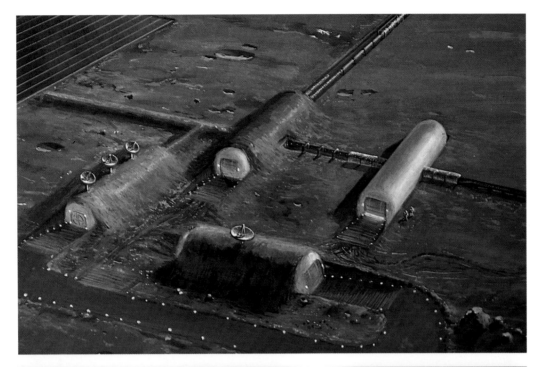

4 The space truck

"This vehicle is performing like a champ. I've got a super spaceship under me."

Robert Crippen, pilot of Columbia *on its maiden flight, April 12, 1981.*

During the afternoon and evening of April 11, 1981, and the early hours of the next morning, hundreds of thousands of people began converging on the Kennedy Space Center. They were waiting in great excitement for the launch of a new breed of space machine, a vehicle that would make space flight a more routine, everyday event. It was to carry the first Americans into orbit since 1975.

The vehicle they had come to see blast off was the Space Shuttle. Developed at a cost of some $10 billion, it was already two years behind schedule. And two days before, its flight had to be postponed nine minutes before lift-off because of technical trouble. Everyone was asking the same question: Will it fly this time?

On the launch pad, over 3 miles (5 kilometers) away from the nearest spectator, the Shuttle was bathed in the brighter-than-daylight glare of powerful floodlights. It was standing upright, bolted down to its launch platform. The main part was a winged vehicle that looked rather like a stubby, medium-sized airliner. This was the orbiter, named *Columbia*. It was mounted on a huge tank, which carried fuel for its engines. Attached to the sides of the tank were two booster rockets.

At about 5 A.M. local time, astronauts John Young and Robert Crippen climbed into the cockpit. Two hours later, at 7 A.M. precisely, they were thundering into the heavens, urged on by the hearts and minds of more than a million spectators. The lift-off was perfect. As *Columbia* sped skywards, it shed its booster rockets and then its fuel tank. Fifteen minutes after lift-off it was cruising in orbit. Three days later, after a flawless mission, it glided back to Earth to make a perfect touchdown at Edwards Air Force Base in California. Exactly seven months later, on November 12, *Columbia* was climbing back into orbit. It was the first time any spaceship had done that.

The Space Shuttle system ushered in a revolution in space flight because it reuses most of the flight hardware. Previously, all space launches had been made by one-off, expendable launch vehicles, such as the Saturn V. The key element in the Shuttle system, the manned orbiter, is a masterpiece of aeronautical and astronautical technology. It was and is the first true aerospace vehicle, able to fly in space and in the atmosphere with equal ease. With a cargo bay big enough to carry a railroad car, it is well nicknamed "the space truck".

Birth of the Shuttle

The Kennedy-inspired marathon to the Moon in the 1960s consumed all of NASA's resources and effectively inhibited future planning for most of the decade. Space scientists knew in their heart of hearts that, in the long term, the Apollo spacecraft and the Saturn V that launched it were Space Age dinosaurs that would die out rapidly when the political and economic climate changed.

And what a dinosaur the Apollo/Saturn V launch vehicle was, measuring 363 feet (111 meters) long and weighing nearly 3,000 tons. It could be used only once. Its first two stages, discarded during launch, ended up in pieces in the Atlantic Ocean. Its third stage, and the Apollo lunar module, usually ended up on the Moon. The Apollo service module burned up during re-entry when the crew came back to Earth. Of the huge rocket that streaked from the launch pad, only the 12-foot (3.6-meter) high command module carrying the crew returned intact to Earth, and that could not be used again. The Apollo series epitomized an extravagant, throwaway policy.

Low-cost launchings

So it really came as no surprise in February, 1967, when the Science Advisory Committee to the U.S. President recommended: "For the longer range, studies should be made of more economical ferrying systems, presumably involving partial or total recovery and use."

By September, 1969, two months after *Apollo 11*'s triumphant landing on the Moon, NASA was outlining its recommendations for America's future in space to the Space Task Group set up by the President. With characteristic breadth of vision, it recommended the construction of a large space station, a lunar orbiting base, and a manned expedition to Mars — all to be accomplished by the turn of the century!

To support this ambitious expansion of space activities, NASA also identified the need to "develop a low unit-mission-cost transportation system that would make Earth-Moon space easily and economically accessible for exploration, applications, science, and technology research." Wrote George E. Mueller, head of the manned flight program: "No law says space must be expensive."

In March, 1970, President Richard Nixon announced that he had approved studies for the development of a reusable transportation system to shuttle to and fro between Earth and space.

Piggyback rides

NASA was confident that a reusable system was now technically feasible. Great strides had been taken in

X-15 rocket plane
Pilot Neil Armstrong poses beside the ultra-fast X-15 rocket plane, which achieved speeds in excess of 3,500 miles per hour (5,500 km/h). Flight data acquired on X-15 flights aided the design of the Space Shuttle orbiter.

aerospace technology during the 1960s — in astronautics through the Mercury, Gemini, and Apollo projects, and in aeronautics through flights by lifting bodies like the M2 and by those of the legendary X-15 rocket planes flying at speeds approaching Mach 7 (seven times the speed of sound).

First, NASA concentrated on developing a fully reusable shuttle vehicle. It would have two winged stages, each manned by a crew of two. The first stage would carry the second stage piggyback until its fuel ran out. Then the second stage would separate and fire its engines, thrusting it into orbit. The first stage meanwhile would return to the launch site and make a horizontal runway landing. So would the second stage when it returned from orbit. The orbital stage would have a spacious cargo bay that could be fitted out with seats for 12 passengers, so that it could carry replacement crews to the proposed space station.

Slashing the costs

A detailed design study for the development of this fully reusable system identified a major problem — cost. At 1971 values the price tag was at least $10 billion. Predictably, this was more than the government would bear, and NASA had to go back to the drawing board to see where savings could be made without sacrificing its overall aims.

A saving of nearly 50 percent was achieved at a stroke by discarding the idea of two fully reusable craft in favor of a single reusable orbiter with reusable boosters. Also, the orbiter was slimmed down by placing the propellants for its engines in a separate, jettisonable fuel tank. With this and other modifications, the cost of development was slashed to about $5.5 billion.

The new concept and costings were acceptable to the Administration, and on January 5, 1972 President Nixon announced the birth of the Space Shuttle. He recommended it to the nation in these terms: "The United States should proceed at once with the development of an entirely new type of space transportation system designed to help transform the space frontier of the 1970s into familiar territory.... It will revolutionize transportation into near space by routinizing it.... It will take the astronomical costs out of astronautics.... This is why commitment to the Space Shuttle program is the right next step for America to take."

Awarding the contract

One final design decision still had to be made. Was the Shuttle to have boosters fuelled by liquid propellants or solids? In the event, solids were chosen because they are of simpler construction and more robust. The only

drawback was that solid rockets had never been used for manned space flights before. Would they be reliable enough? Some authorities, notably Wernher von Braun, believed that solids should never be used for space flight. In the light of the *Challenger* disaster (page 105), perhaps they were right.

Nevertheless, solids were chosen, and on March 15, 1972, NASA announced the final Shuttle design — manned orbiter, external tank for its fuel, and twin solid rocket boosters. That July it awarded North American Rockwell Corporation the contract for developing and building the orbiter. Within a year detailed design studies were nearly complete. It was anticipated that the first orbital flights would take place as early as 1979.

As history records, this was wildly optimistic in view of the host of the new technologies that had to be developed. In the event, the major hold-ups were due to problems with the totally new main engines and the ceramic tiles of the orbiter's heat shield.

Prototype *Enterprise*
The first orbiter to be built was a prototype, destined never to leave Earth. It was essentially a body shell that could be used for flight tests in the atmosphere to verify aerodynamic design. NASA's preferred name for the prototype orbiter was *Constitution*. But a campaign by fans of the cult sci-fi TV series "Star Trek" persuaded President Ford to name the prototype *Enterprise*. The four operational orbiters were to named *Columbia*, *Challenger*, *Discovery*, and *Atlantis*, after famous naval ships.

Enterprise had its few brief moments of glory in 1977 when it took part in approach and landing tests at Edwards Air Force Base in California. It was carried piggyback on a converted Boeing 747 jet and then released, gliding unpowered to a runway landing. The flights were flawless, the low-speed aerodynamic characteristics of the orbiter perfect.

How the orbiter would fare when re-entering the Earth's atmosphere at 25 times the speed of sound was a different matter. How it would behave as a spacecraft was also an unknown quantity. No one would know until the first operational orbiter *Columbia* took to the skies. That was still four years away.

Flying free
Prototype orbiter *Enterprise* lifts away from the 747 carrier plane during the second free flight in the ALT (approach and landing test) program on September 13, 1977. At the controls are Joe Engle and Richard Truly.

▲
Orbiter mating
The Shuttle stack is put together in the cavernous Vehicle Assembly Building. The solid rocket boosters are assembled first. Then the external tank is placed in position. Finally the orbiter is lowered onto the tank, and attached and plumbed to it.

Lift-off!
Atlantis streaks from the launch pad on October 30, 1985, on its maiden flight (mission 51-J). The dramatic camera angle emphasizes the formidable power of its twin solid rocket boosters. Contrast the columns of flame and smoke from the boosters with the transparent exhaust gases from the three main engines.

A super spaceship
Columbia punched its way into the Florida skies on April 12, 1981. A textbook lift-off was followed by a trouble-free two days in space, a flawless re-entry, and a perfect touchdown. During the mission, pilot Robert Crippen enthused: "This vehicle is performing like a champ. I've got a super spaceship under me." For Commander John Young, *Columbia* was "a magnificent flying machine".

The pioneering mission of *Columbia* was designated STS-1, the STS standing for Space Transportation System. The first nine flights of the Shuttle were designated with STS numbers. The next 16 were given a designation that baffled practically everyone! Then with *Discovery*'s flight (STS-26) in 1988, the first after the *Challenger* disaster, the designations reverted to STS numbers once again.

Orbiter and ET
The orbiter measures over 122 feet (37 meters) long and has a wingspan of nearly 78 feet (24 meters). It is part rocket, part spacecraft, part plane. It takes off vertically like a rocket, maneuvers in space by means of little rocket thrusters like a spacecraft, and then returns to Earth like a plane, using its aerodynamic surfaces — wings and tail — to maneuver in the air. It lands horizontally on a runway like a plane, or rather like a glider, because its descent is unpowered.

On the launch pad the orbiter is mounted on the external tank (ET), which holds propellants for its engines. The ET is the largest piece of Shuttle hardware, measuring 154 feet (47 meters) long. It houses two tanks, one containing liquid hydrogen fuel and the other liquid oxygen oxidizer. When full, it can hold over 450,000 gallons (2 million liters) of propellants. The propellants are cryogenic, or supercold, so the ET is sprayed with a thick layer of orange insulating foam as protection against frictional heating as the Shuttle ascends through the atmosphere.

The solid rocket boosters
The ET in effect serves as a backbone to the whole Shuttle cluster. Also attached to it are the twin booster rockets, the solid rocket boosters (SRBs). The SRBs, which measure 149 feet (45.5 meters) long and nearly 13 feet (4 meters) in diameter, are made of high strength steel ½-inch (1.3 centimeters) thick. They are constructed of 11 segments, held together by steel pins.

The joints are sealed by so-called O-rings, designed to prevent the escape of gas. It was the failure of an O-ring that led to the tragedy of *Challenger* on the Shuttle's twenty-fifth flight in 1986. The solid propellant packed into the SRBs is made up of finely powdered aluminum, which serves as fuel, and ammonium perchlorate as the oxidizer. They are bound together by a kind of synthetic rubber.

The SRBs fire a few seconds after the orbiter's main engines, and lift-off occurs a fraction of a second later. In only two minutes the SRBs run out of fuel. But by then their work has been done — they have boosted the Shuttle through the thickest part of the atmosphere, where there is maximum drag, or air resistance.

They separate from the rest of the Shuttle stack and plunge back to Earth. At an altitude of about 16,000 feet (5,000 meters), small drogue parachutes deploy from their nose cones to slow down their descent. At 6,500 feet (2,000 meters) three main parachutes open to lower the SRBs gently into the sea, where recovery ships are standing by. The ships tow them back to the Kennedy launch site for refurbishment and subsequent reuse.

Main and auxiliary engines
The three main engines of the orbiter are among the most efficient engines ever produced. Although they weigh only about 6,500 pounds (3,000 kilograms), they put out a thrust of up to 450,000 pounds (200,000 kilograms). They consume the propellants at the rate of nearly 63,000 gallons (300,000 liters) per minute. If they drew the water from a family-sized swimming pool at the same rate, they would empty it in 30 seconds!

The main engines fire from about 7 seconds before

lift-off for about 8½ minutes, until the ET begins to run dry. They then cut out, and the ET is jettisoned. It falls back and breaks up in the ocean, the only item of Shuttle hardware that cannot be used again.

About two minutes after main engine cut-off and ET separation, two smaller engines alongside the main ones in the tail engine pod fire and boost the orbiter to its orbital velocity of about 17,500 miles per hour (28,000 km/h). These engines are known as the orbital maneuvering system (OMS) engines. They use an interesting cocktail of propellants, hydrazine and nitrogen tetroxide, which burst spontaneously into flame when they are mixed. The OMS engines can also be fired to change the orbit of the orbiter. And at the end of the mission they are fired for the de-orbit burn, the retrobraking maneuver that reduces the speed of the orbiter to below orbital velocity, so that gravity can pluck the craft from space.

While the orbiter is in space, it relies on thruster rockets to maintain and change its attitude, or position. These thruster rockets make up what is called the reaction control system (RCS). There are 44 thrusters in all, housed in clusters in the nose and the tail pod. They use the same kind of propellants as the OMS.

Orbiter construction

The orbiter is built up of six main units — forward fuselage, wings, mid-fuselage, payload-bay doors, aft fuselage, and tail. The construction of the airframe is surprisingly conventional, following established aeronautics practice. It is made of ribs and spars in high-strength aluminum alloys, with titanium in places that experience particularly high stress. Since aluminum softens at elevated temperatures, the airframe has to be protected with an effective heat shield.

The primary heat shield material is a silica ceramic applied in the form of thick tiles. On *Columbia*, for example, there are more than 30,000 tiles. They are thickest on the orbiter's underside, which bears the brunt of re-entry heating. The nose and leading edges of the wings experience the maximum temperatures, up to 2,700°F (1,500°C), and are covered by an even more effective carbon resin insulation. On less critical areas, such as the upper fuselage, insulating felt affords adequate protection.

On the flight deck

The nerve center of the orbiter is the flight deck. This is located on the upper deck level of the pressurized forward fuselage. Underneath it is the mid-deck section that serves as the living quarters of the astronauts.

At the front of the flight deck is the cockpit, from where the mission commander and pilot fly the machine.

The cockpit looks much the same as that of a modern airliner such as the Boeing 747-400 or the European Airbus 320. It is crammed with arrays of switches, buttons, keyboards, and instruments. Many of the instruments, such as the artificial horizon and gyrocompass, would be familiar to airline pilots.

The most noticeable difference between the flight deck console of the orbiter and that of an airliner is the presence of three cathode-ray tube (CRT) displays. These are linked to the orbiter's computer system and can display all kinds of flight and engineering data.

The computer system is vital to the operation of the orbiter, which is far too complicated a machine to be flown manually. At peak periods of activity, such as lift-off, it has to perform over 325,000 operations every second. No human pilot or even a team of pilots could cope with that kind of workload. The orbiter flies automatically under computer control for much of the time. And even when pilot and commander do take over the controls, all operations are channelled through the computer system.

The computer system is so essential that it is made "quad-redundant". In other words, there are four main computers that function all of the time, with a fifth as a back-up. The four primary computers process the same information. If one of the computers disagrees with the others, it is overruled. If none of them can agree, the back-up takes over.

Landing approach
After successfully deploying the planetary probe *Magellan* (mission STS-30), *Atlantis* drops down to the runway at Edwards Air Force Base on May 8, 1989. This view shows the dark heat-shield tiles that minutes earlier would have been glowing red-hot as the orbiter re-entered the atmosphere at 25 times the speed of sound.

Booster separation
Early concepts for the Space Shuttle concentrate on a fully reusable winged booster and winged orbiter system. Both are to be manned. North American Rockwell come up with this splendid design in 1970 for a delta-winged orbiter. It is to be lifted piggyback high into the atmosphere by a booster, seen here peeling away to return to Earth.

Shuttle lift-off
This alternative 1971 design for a fully reusable shuttle craft has a more conventional look about it. The problem with all the fully reusable concepts is their cost — up to $10,000 million at 1970 values. Two advanced-technology aerospace craft would have to be built, with long-life components for repeated use.

Orbiter re-entry
North American Rockwell,
which eventually receives the
contract for the current Space
Shuttle, carries out another
design study in 1970 for a fully
reusable system. In this design
the orbiter has stub wings, is
183 feet (56 meters) long, and
has a wingspan of 125 feet (38
meters).

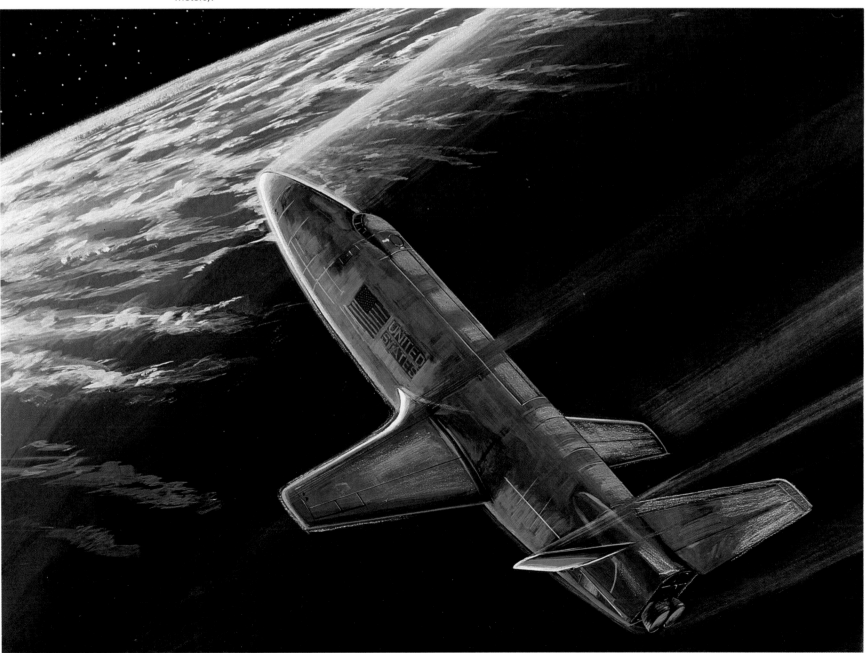

Blast-off
By 1972 the fully reusable concept has been abandoned due to excessive cost, and the present system of a smaller winged orbiter, twin rocket boosters, and an external propellant tank is approved. The orbiter, in this concept by Grumman Aerospace, looks more like the modern version.

▶ **Boosters away**
After the boosters have done their job of thrusting the Shuttle through the denser part of the atmosphere, they fall away.

▶ **Booster recovery**
Parachutes lower the boosters to a low-speed splashdown at sea. They are then towed back to the launch site for overhaul and subsequent reuse.

External tank jettison
By 1976, artists' concepts of
Space Shuttle operations are
featuring the modern version
of the orbiter. Here it is
jettisoning the external tank.
The main engines have
stopped firing of course, for
they draw their propellants
from the tank. Soon the two
smaller orbital maneuvering
system engines will fire to insert
the craft into orbit.

2·1·77

◀

Enterprise hits the road
On the last day of January,
1977, prototype orbiter
Enterprise journeys by road
from North American
Rockwell's plant at Palmdale,
where it was built, to the
Dryden Flight Research
Center, where it will undergo
flight tests in the atmosphere.
This painting is by Robert
McCall.

Enterprise on a jumbo
At NASA's Dryden Research
Center, _Enterprise_ takes part in
a series of approach and
landing tests to verify its low-
speed aerodynamic
performance. It is carried into
the air by a converted Boeing
747 jumbo jet, as depicted in
this artist's impression. In the
first tests the orbiter remains
fixed to the jumbo carrier.

"Touchdown with Chase Planes"
After eight captive flights mounted on the carrier jet, prototype orbiter *Enterprise* begins a series of free flights, commencing on August 12 and ending on October 26, 1977. The orbiter is released at an altitude of about 23,000 feet (7,000 meters). It takes about five minutes to glide to a lake bed landing, accompanied by chase planes, as in this painting by Tom O'Hara.

Tom O'Hara

***Enterprise* roll-out**
On May 1, 1979, *Enterprise* is rolled out of the Vehicle Assembly Building at the Kennedy Space Center, and integrated with solid rocket boosters and external tank into a fully assembled Shuttle stack. It provides a foretaste of the upcoming Shuttle era. Nicholas Solovioff's painting shows the stack moving at a snail's pace along the crawlerway towards the launch pad, visible on the skyline over 3 miles (5 kilometers) away.

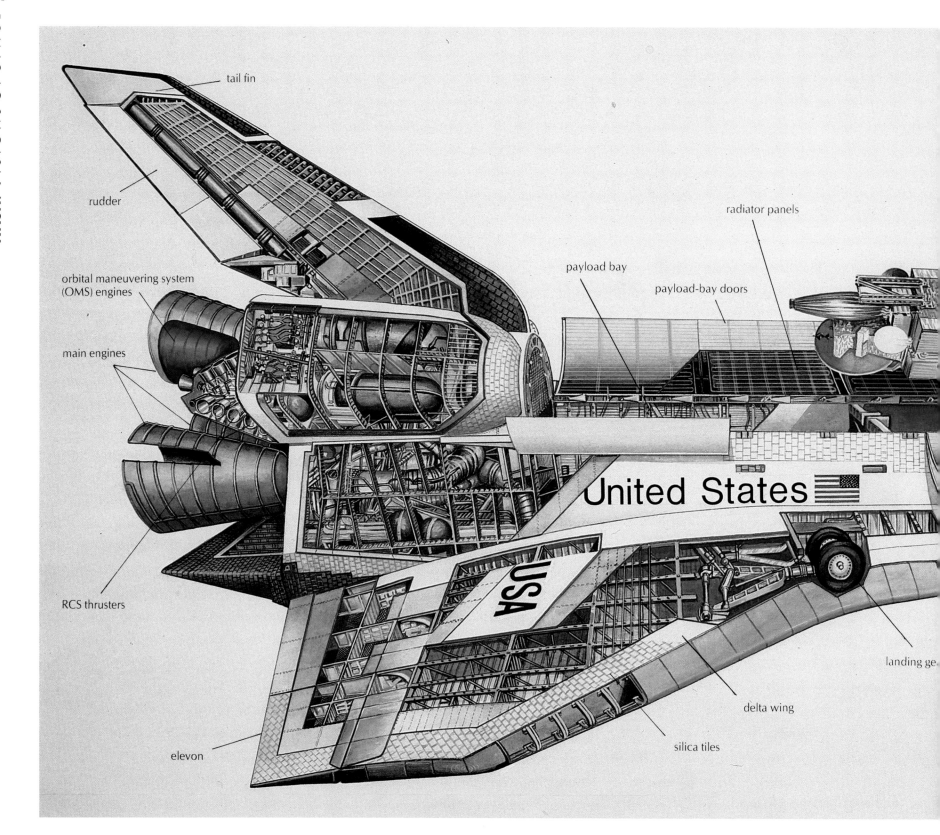

tail fin

rudder

orbital maneuvering system
(OMS) engines

main engines

RCS thrusters

elevon

radiator panels

payload bay

payload-bay doors

United States

USA

landing ge

delta wing

silica tiles

Orbiter cutaway

This artwork of the delta-winged Space Shuttle orbiter shows details of construction and the location of major components. There are three main engines and two orbital maneuvering system (OMS) engines in the tail pod. The pod also houses two sets of thrusters which are part of the reaction control system (RCS). The RCS thrusters are fired in orbit to alter the attitude of the orbiter. There are further sets of RCS thrusters in the nose. Overall, the orbiter measures a little over 122 feet (37 meters) long and has a wingspan of 78 feet (23.8 meters).

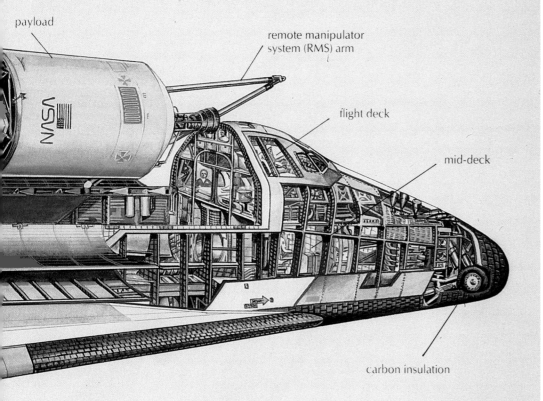

payload

remote manipulator system (RMS) arm

flight deck

mid-deck

carbon insulation

The Shuttle stack

The Shuttle remains in this configuration for just 2 minutes after lift-off. Then the solid rocket boosters separate, followed some 6 minutes later by the external tank. Only the orbiter reaches orbit. From the tip of the external tank to the base of the booster nozzles, the Shuttle stack measures a little over 184 feet (56 meters).

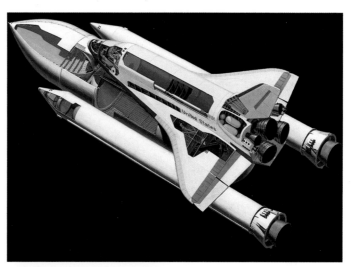

Red-hot grip

The most important parts of the orbiter's heat shield, or thermal protection system (TPS), are the silica tiles, typically about 6 inches (15 centimeters) square. The tiles have the most incredible heat-insulating properties. A block of tile material, red-hot inside, can be cool enough outside to be picked up, as this photograph demonstrates.

◀◀
Big boosters
The Shuttle solid rocket booster (SRB) is the largest rocket of its type ever flown and the first to be designed for recovery and reuse. This drawing by Maria Epes records a scene during SRB recovery tests in July, 1980. The rear of an SRB is shown in the right foreground, towering above the technician.

All systems go
At lift-off all five Shuttle rockets fire to generate enough thrust to lift the 2,000-ton Shuttle stack into the heavens. This pre-Shuttle era artwork contrasts the filthy exhaust belching from the SRBs with the pale blue flames issuing from the main engines. The main engines are environmentally friendly, for they combine hydrogen and oxygen to form water vapor.

On the flight deck
In their flight positions in
Columbia, STS-1 commander
John Young (left) and pilot
Robert Crippen run through
check lists during a rehearsal
for the first Shuttle mission. The
orbiter's flight deck is not
unlike that of a modern
airliner, but the
instrumentation is more
extensive and more complex.
The forward flight console also
includes three cathode-ray
tubes, on which the crew can
display a variety of data.

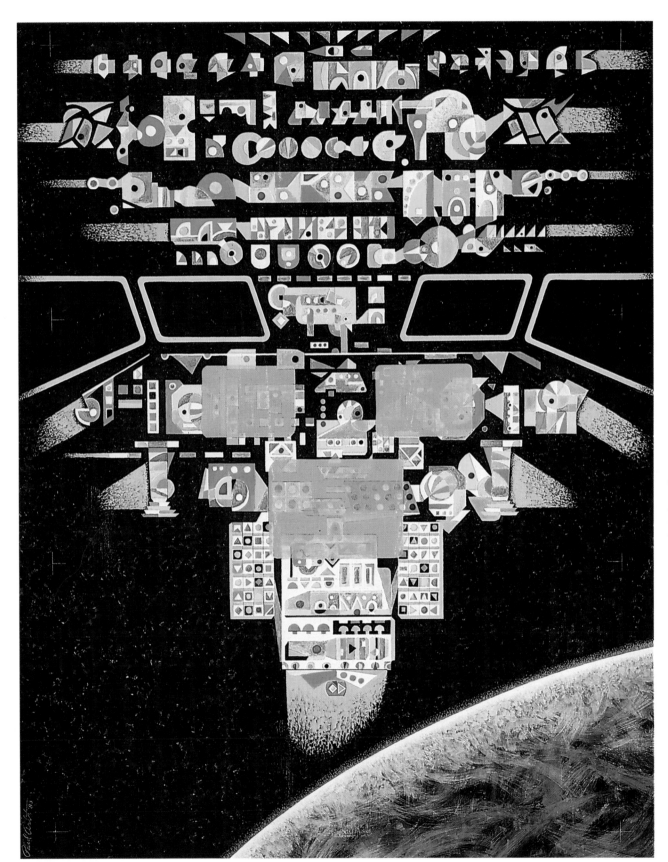

"Flight Deck Fantasy"
The bewildering complexity —
to the lay person at least — of
the Shuttle flight deck inspired
this work by Paul Arlt, one of
the artists invited to cover the
STS-6 mission, the first flight of
orbiter *Challenger*.

5 Shuttle missions

"The fourth landing of the *Columbia* is the historical equivalent of the driving of the golden spike which completed the first transcontinental railroad. It marks our entrance into a new era."

President Ronald Reagan at the conclusion of STS-4, the final test flight of the Space Shuttle, July 4, 1982.

In November, 1981, Space Shuttle orbiter *Columbia* blasted off the launch pad at the Kennedy Space Center for an unprecedented second journey into space. This flight, like the first one seven months earlier, went like a dream, as did *Columbia's* next two test flights in 1982. After the fourth flight, in July, the Space Shuttle was pronounced fully operational. And in November *Columbia* at last began to earn money for NASA, by launching two commercial communications satellites.

A second orbiter, *Challenger*, made its space debut in April, 1983. That year saw many successful satellite launchings from the Shuttle, the testing of the Shuttle spacesuit, the flight of the first American woman in space, Sally Ride, and the launch of Spacelab.

However, it was not until 1984 that the extraordinary versatility of the Shuttle and its astronauts really became apparent. The year saw Bruce McCandless become the first astronaut to make an untethered spacewalk. It saw other Shuttle astronauts capture, repair, or recover three satellites during the most spellbinding spacewalks of the Space Age. The third Shuttle orbiter, *Discovery*, became operational in August.

In 1985 there was more spacewalking to fix a dead satellite and practice space construction techniques. In October the fourth orbiter, *Atlantis*, took to the skies. This brought the shuttle fleet to its full complement of four. A second flight by *Atlantis* in November, after an exceptionally quick turnaround, ended a year in which a record nine Shuttle flights had taken place. It seemed as if the promise that the Shuttle would transform space flight into an everyday, routine operation was coming to pass.

But on January 28, 1986, on a clear and particularly frosty morning at Cape Canaveral, the illusion of the ordinariness of space flight was shattered at a stroke. Seventy-three seconds into the Shuttle program's twenty-fifth and *Challenger's* tenth mission, the accelerating Shuttle stack erupted into a mighty orange fireball. *Challenger* was blasted apart and its seven-member crew perished.

It took 32 months of painstaking investigation, agonizing, and system modifications to get the Shuttle up and running again. Not until September, 1988, did American astronauts return to space. Since then, with a restructured management and tighter operational procedures, the impetus and optimism of the Shuttle era has been regained.

Mission astronauts

The days are long gone when astronauts needed the elusive quality of daredevilry and machismo that came to be called the "Right Stuff". Nor do they need the ability to "fly by the seat of their pants" like the early pioneers. Flying the Shuttle is much like flying in a very sophisticated airliner.

Gone are the body-crushing accelerations as Shuttle crews ascend into orbit. These are now a modest 3g (three times the pull of gravity), comparable with those we Earth-bound mortals can experience in some of the more terrifying roller-coaster rides. Gone also are the spacesuits the early astronauts wore all the time. The Shuttle orbiter maintains a comfortable shirt-sleeve environment. Only if astronauts need to step outside their craft are spacesuits required.

As a result, anyone who is reasonably fit can now fly into space aboard the Shuttle with the minimum of training. This makes it possible for people to be launched into space on a one-time basis, for example, a scientist to carry out a specific experiment. Such scientists, known as payload specialists, do not form part of NASA's corps of professional astronauts.

Among the "pros" there are two categories, pilot-astronauts and mission specialists. The pilot-astronauts (commander and pilot) are more like the astronauts of earlier eras, being highly skilled fliers with thousands of hours of high-speed jet flight. The mission specialists do not need to fly, since their task on board the Shuttle is to carry out mission requirements, such as launching satellites and supervising experiments.

Mission training

The astronauts begin training for a particular Shuttle flight months before the scheduled lift-off. The headquarters for astronaut training is the Johnson Space Center at Houston, Texas. The two categories of astronauts sometimes train separately, sometimes together. The pilot-astronauts, for example, put in a lot of flying to maintain their piloting skills during the months between space flights. They also train in the Shuttle mission simulator, which exactly mimics the

***Challenger* roll-out**
Orbiter *Challenger* emerges from the Vehicle Assembly Building in mid-October, 1985, mated with external tank and solid rocket boosters in its launch configuration. The Shuttle stack is beginning its slow 3.5-mile (5.5-kilometer) journey to the launch pad. It is carried by one of the world's largest land vehicles, the eight-tracked crawler transporter, which is 135 feet (40 meters) long and nearly 115 feet (35 meters) wide.

operation of the real craft in flight.

The mission specialists often join them in simulations of lift-off and other flight operations. They separately take part in other simulations for their own particular role in space, for example, satellite launching or spacecraft capture and retrieval. Many of them undergo training for spacewalking in a huge water tank called the neutral buoyancy chamber or water-immersion facility. In the water tank they wear a modified spacesuit weighted to give neutral buoyancy, in which state they neither rise nor sink. It is the nearest astronauts can get on Earth to the weightlessness they will experience in space.

Mission countdown

The prime launch site of the Shuttle is Complex 39 at the Kennedy Space Center, built originally for launching the Apollo/Saturn Moon rockets. Lift-off may take place from one of two launch pads, 39A or 39B. The Shuttle stack, comprising orbiter, solid rocket boosters, and external tank, is put together on a mobile launch platform in the Vehicle Assembly Building (VAB). It is "rolled out" to the launch pad on top of one of the giant crawler transporters, which were also built for Apollo. Roll-out usually takes place three weeks before the scheduled date of lift-off.

About three days before launch, the countdown begins — the counting down of time to the precise moment of lift-off, T. The large countdown clock at the press site that looks towards the launch pad is started. This shows less than the time remaining because it makes allowance for built-in holds. These are periods when the countdown is stopped deliberately for minutes or hours. This allows mission managers the opportunity to review progress and for engineers to catch up with launch preparations, should they fall behind.

About two hours before the scheduled lift-off the crew members enter the orbiter and strap themselves in. The external tank is topped up with propellants. The orbiter is powered up. All systems are go. At the built-in hold of T minus 9 minutes there is a final status check when the launch director consults with a mission management team. If all are in agreement that the launch should take place, the launch director decides it is "Go for launch". If there are serious reservations — on the state of the weather, for example — the launch will be scrubbed.

This last-minute consultation with a team drawn from Shuttle contractors and NASA managers was instituted following the *Challenger* disaster. On that occasion serious reservations about the advisability of launching were not passed on to the persons who had their fingers on the firing button.

Mission Control

During the countdown and at lift-off the Shuttle is under the control of the Launch Control Center at Complex 39, which is located on the launch-pad side of the VAB. Once the Shuttle has lifted-off, however, control is switched to Mission Control at the Johnson Space Center, which has the famous callsign "Houston".

In the mission operations control room at the Johnson Space Center, flight controllers sit at instrument-packed consoles, each with responsibility for monitoring particular activities happening on the Shuttle. At the front of the room are large video screens on which data and TV transmissions can be projected. There is also a moving display showing the location of the Shuttle at any time on a map of the world.

In overall charge is the Flight Director. He communicates with the astronauts, not directly, but through CapCom, the capsule communicator, who is also an astronaut. The title is a throwback to the early days of space flight when astronauts were lofted into space in cramped capsules.

Mission operations

It takes the orbiter less than an hour to reach its desired orbit, typically 190–220 miles (300–350 kilometers) above the Earth. One of the first tasks of the crew in orbit is to open the doors of the payload bay. This is necessary to expose the radiators on the underside of the doors. These get rid of excess heat that has built up in the orbiter during ascent.

Whatever else it may be used for, the orbiter is first and foremost a cargo carrier — a space truck. And its usual cargo is one or more satellites. These are carried in the cavernous payload bay, which measures some 60 feet (18 meters) long and 15 feet (4.6 meters) across.

Most of the satellites the shuttle carries are comsats — communication satellites. They need to be launched into geostationary orbit 22,300 miles (36,000 kilometers) above the Earth. This altitude is far beyond the capability of the Shuttle, so they are first launched into low orbit and then boosted higher by an attached rocket stage. Many use a booster called PAM, the payload assist module. For heavier payloads, such as TDRS, a more powerful unit called the IUS (inertial upper stage) is employed.

The TDRS (pronounced "teedris") tracking and data relay satellites now play a central role in Shuttle and satellite space-Earth communications. Spaced strategically around the globe in geostationary orbit, they act as relay stations to pass on communications and telemetry signals between spacecraft and ground control. In particular they enable continuous communications to be maintained with the Shuttle .

TDRS deployment
The tracking and data relay satellite, TDRS, emerges from *Discovery*'s payload bay in September, 1988. When *Discovery* has backed off to a safe distance, the attached booster rocket will fire to lift the satellite into an orbit 22,300 miles (36,000 kilometers) above the equator.

Remote manipulation

Satellites and other payloads are launched from the Shuttle by remote control by mission specialist astronauts working at the aft crew station on the upper deck of the orbiter. At the aft crew station there are two windows that look into the open payload bay and another two overhead in the roof. These allow the astronauts good vision during satellite-launching operations. Many PAM-boosted satellites are launched from cradles in the payload bay. They are set spinning and then sprung out. Other satellites are simply spun out of the bay frisbee-style.

Most of the large payloads, however, are plucked and placed — plucked out of the bay with the Shuttle "crane" and literally placed into orbit. This method was used, for example, to launch the huge passive satellite called the Long Duration Exposure Facility (LDEF) and the Hubble Space Telescope.

The Shuttle crane, correctly termed the remote manipulator system (RMS), is a multi-jointed robot arm some 50 feet (15 meters) long. It is controlled from the aft crew station by two hand-controllers. It carries a closed-circuit TV camera, which relays pictures to monitor screens at the station. The RMS also plays a key role in satellite retrieval operations, with or without the help of spacewalking astronauts.

Going EVA

Satellite retrieval, repair, and recovery operations have become an integral part of Shuttle activities. Despite having to work in bulky spacesuits, spacewalking astronauts have been able to carry out quite intricate repairs to malfunctioning satellites. The correct term for spacewalking is EVA, or extravehicular activity, and the Shuttle spacesuit is termed the EMU, or extravehicular mobility unit.

The EMU is a multilayer garment modeled on the spacesuit the Apollo astronauts wore on the Moon. It comes in two parts — upper and lower torso. The upper torso carries the life-support backpack, which supplies the astronaut with oxygen and cooling water for the water-cooled long johns he or she wears next to the skin. The lower torso, or trousers, joins the upper at the waist with an airtight seal.

For moving around in space the astronauts use a jet-propelled backpack, called the MMU, or manned maneuvering unit. It is a kind of flying chair, propelled by sets of tiny thrusters — 24 in all — which spurt out jets of compressed nitrogen gas. The astronaut drives and steers the MMU by firing sets of thrusters in the appropriate direction using hand controls on the armrests.

Mission highlights

As Neil McAleer wrote in the run-up to the flight of STS-26, the first post-*Challenger* mission: "There are enough Space Age firsts and records established during the first 24 Shuttle flights to fill an almanac."

Dozens of commercial scientific and military satellites were delivered into orbit; two were retrieved and brought back to Earth; and two were plucked out of orbit, repaired in the Shuttle's payload bay, and returned to orbit. Four flights of the European-built *Spacelab* took place, bringing non-Americans into the space program for the first time. Each of these flights accumulated enough data, if converted into words, to fill thousands of sets of *Encyclopedia Britannica*.

In the post-*Challenger* era, Shuttle missions continued to excite the imagination. Among the highlights have been the launch of two probes which herald a new era in planetary space science, *Magellan* to Venus and *Galileo* to Jupiter. Astronomically, the highpoint has been the launch of the much-delayed Hubble Space Telescope, designed to provide a new and crystal-clear window on the universe. The capture and return to Earth of the LDEF in an 11-day mission, the longest ever, again demonstrated the versatility of the Shuttle and the astronaut corps. Such versatility will be needed in abundance in the coming years when the Shuttle enters the next challenging phase of space exploration — the construction in orbit of the space station *Freedom*.

Some of the highlights of Shuttle missions from STS-1 through STS-3l in April, 1990, are given in the next few pages. The names of the crew members and the launch dates are given in the general Manned Mission Chronology on page 162. All launchings took place at the Kennedy Space Center in Florida, and the landings were at Edwards Air Force Base in California, except where otherwise stated.

Inspecting Solar Max
It is George Nelson's turn to ride the cherry picker on the following mission (41-C) in April, 1984. He is inspecting the newly captured satellite Solar Max, which he and colleague James Van Hoften will later mend and restore to working order.

▶
Weightless thingumajig
Mission 41-G, in October, 1984, sees two female astronauts ride into space together for the first time. Here off-duty, Kathryn Sullivan (left) and Sally Ride seem delighted with the "artistic" thingumajig they have made from bits and pieces they have found floating around.

◀◀
On the cherry picker
On mission 41-B, in February, 1984, Bruce McCandless uses the Shuttle's remote manipulator system (RMS) arm like a cherry picker. He is standing on a small platform called the mobile foot restraint.

The proving flights

Beginning on April 12, 1981, orbiter *Columbia* flies into space five times in 21 months, proving the reusable concept of the Shuttle system. Minor problems abound, including computer problems on STS-1, faulty APUs (auxiliary power units) on STS-2, and loss of the SRBs (solid rocket boosters) on STS-4. The eight-day mission of STS-3 ends at the back-up landing site at White Sands, New Mexico, because of flooding at Edwards Air Force Base. The first commercial flight of the Shuttle, STS-5, which sees the launch of two communications satellites, also sets a record by carrying four astronauts into space for the first time.

Challenger's debut year

Problems with the main engines and TDRS payload delay *Challenger's* maiden flight from January 20 until April 4, 1983. On its first mission (STS-6) Donald Peterson and Story Musgrave test the new Shuttle EMU, or spacesuit, for the first time. On its second (STS-7), Robert Crippen becomes the first astronaut to make two Shuttle flights. The flight is also marked by Sally Ride becoming the first American woman to fly in space. The next launch (STS-8) sees *Challenger* making the first nighttime Shuttle lift-off.

Columbia returns to space on November 28, 1983, carrying the European-built Spacelab into orbit for the first time (see page 127).

Into the Buck Rogers era

The baffling mission designations begin! Bruce McCandless on mission 41-B test-flies the MMU (manned maneuvering unit) on February 7, 1984, making the first untethered spacewalk. On the same mission two satellites, Westar VI and Palapa B, are deployed, but their boosters refuse to ignite and they remain in low, useless orbits. *Challenger* makes the first landing at the Kennedy Space Center.

George Nelson flies the MMU on the next flight (41-C) in a vain attempt to capture the ailing satellite Solar Max. Later the shuttle's robot arm succeeds and drops the satellite into *Challenger*'s payload bay. There, Nelson and James van Hoften carry out repairs to the electronic systems. Then, working again, Solar Max is placed back in orbit.

The next flight (41-D) sees the debut on August 30, 1984, of the third orbiter *Discovery*, much delayed by a computer malfunction on June 25, an engine cut-off the next day a few seconds before lift-off, and a further computer anomaly on August 28. Five weeks later, on mission 41-G, *Challenger* carries the largest crew ever into orbit (seven), including for the first time two women, Sally Ride and Kathryn Sullivan. The latter becomes the first U.S. woman astronaut to make a spacewalk. Robert Crippen makes his fourth Shuttle flight and brings in *Challenger* for the second landing at Kennedy.

Riding the stinger

Discovery's second mission (51-A) is one of the most ambitious ever, seeking to recover the errant satellites lost in low orbit on 41-B. Employing a unique kind of docking probe called the stinger, Dale Gardner and Joseph Allen in MMUs capture the satellites and stow them in the payload bay for return to Earth in the most daring spacewalks of the Space Age.

Little is known of *Discovery's* next mission (51-C), dedicated to the Department of Defense (DOD). But the one following (51-D) attracts more attention than usual due to the presence on board of U.S. Senator Jake Garn. Embarrassingly in the circumstances, one of the satellites launched, Leasat 3, fails to activate after deployment, and despite attempts to flick a switch with a makeshift "flyswatter" Leasat stays dead. Further indignity is added when *Discovery* blows a tire on landing at Kennedy.

The pace hots up

Challenger lifts off only 17 days after *Discovery*, making this the fastest launch rate yet. On this mission (51-B) it carries *Spacelab 3* into orbit, with two monkeys and 24 rats aboard. A routine satellite-launching mission (51-G) by *Discovery* occurs in June, followed by another Spacelab trip (*Spacelab 2*, 51-F) a month later. A near-emergency situation develops on lift-off when one main engine shuts down.

Spectacular spacewalking to recover and repair the rogue satellite Leasat from 51-D marks *Discovery's* sixth mission (51-I). Honors fall to William Fisher and

Capturing the rogue
Flying the manned maneuvering unit (MMU) in November, 1984, mission 51-A spacewalker Dale Gardner attaches a device called a stinger to the Westar VI satellite. Using the MMU's thrusters, Gardner will next propel the satellite back to *Discovery* and, with the aid of Joseph Allen, stow it in the payload bay.

James "Ox" van Hoften for immaculate repair work during EVAs totalling nearly 12 hours. A hush-hush DOD mission (51-J) comes next when the fourth orbiter *Atlantis* makes a flawless maiden flight. *Challenger's* last full mission (61-A) is a Spacelab flight (*Spacelab D1*) dedicated to West Germany. The largest crew ever, eight, fly into space together. The normal maximum crew of seven fly on *Atlantis's* second flight, which sees a demonstration by Jerry Ross and Sherwood Spring of techniques of building structures in orbit.

Trouble all the way

On the first mission (61-C) in 1986, the worst year in U.S. space flight history, *Columbia* lifts off on January 12 after a record seven delays because of engine problems and poor weather. A succession of minor problems also delays the next mission (51-L), dubbed "the teacher's flight" because it includes a teacher, Christa McAuliffe, who is scheduled to give lessons from orbit. Orbiter *Challenger* eventually lifts off on January 28 after a bitterly cold and frosty night. It is making its tenth flight.

The seemingly perfect lift-off quickly goes awry. A faulty seal in the right-hand solid rocket booster allows flame to blowtorch one of the struts holding the booster to the external fuel tank. The strut breaks away, the booster swings into the tank, which ruptures and triggers off a mighty explosion at 73 seconds into the mission. *Challenger* is blown apart. In the awesome fireball that results, Richard Scobee, Michael Smith, Judith Resnik, Ellison Onizuka, Ronald McNair, Gregory Jarvis, and Christa McAuliffe meet their death. They are the first Americans to die during a space flight. All further Shuttle flights are suspended.

Rising from the ashes

The Shuttle is grounded for more than two and a half years following the *Challenger* disaster. Over 400 major and minor modifications are made to the solid rocket boosters, the main engines, the external tank, and the orbiter to insure that a similar disaster will not recur. Changes are also made to NASA launch procedures to improve safety. On September 29, 1988, *Discovery* blasts off on a four-day mission (STS-26) to help reestablish the credibility of the Shuttle system. Thankfully it is a perfect flight, on which a TDRS comsat is accurately launched. A little over two months later *Atlantis* departs on a classified DOD mission (STS-27), deploying a spy satellite.

Discovery makes the first and last missions of 1989, carrying another TDRS into orbit on STS-29 (March) and a DOD payload on STS-33 (November), a spectacular night launch. For planetary scientists the two most important payloads of the year are the *Magellan* probe to Venus (STS-30, May) and the *Galileo* probe to Jupiter (STS-34, October). The orbiter both times is *Atlantis*. The launch of *Columbia* in between (STS-28) is another DOD mission.

Columbia also spearheads the busy launch program for 1990, with the successful capture in January by STS-32 of the LDEF, the long duration exposure facility. This has been in orbit since 1984 and would have been retrieved from orbit in 1986 had Shuttle operations not been suspended. This mission sets a record for the longest Shuttle flight to date, nearly 11 days. *Atlantis* is launched in February on yet another DOD mission (STS-36). Astronomers pin their hopes on the launch by *Discovery* in April of the Hubble Space Telescope, the most expensive payload the Shuttle has ever carried ($1.5 billion). It promises the clearest view yet of the heavens and a look back in time to when the universe began.

Seven star voyagers
"Seven star voyagers" is how President Ronald Reagan described the crew who met their deaths in the Florida skies that January day. Said the President: "They slipped the surly bonds of Earth to touch the face of God." They are commemorated in this artwork, which includes the crew patch for the mission (51-L) and portraits of the crew. They are (clockwise from the left): Ronald McNair, Ellison Onizuka, Judith Resnik, Richard Scobee, Michael Smith, Christa McAuliffe, and Gregory Jarvis.

America returns to space
After more than two and a half years in limbo, the Space Shuttle roars from the launch pad on September 29, 1988, on mission STS-26. Four days later Vice-President George Bush is on hand at Edwards Air Force Base to greet the crew when *Discovery* lands after a flawless four-day flight.

Early morning roll out of the Col
Kennedy Space Center.

"*Columbia* roll-out from the VAB"
Two years behind schedule, orbiter *Columbia* emerges from the Vehicle Assembly Building and begins its slow journey to Launch Pad 39A at the Kennedy Space Center. The date is December 29, 1980. The lift-off of *Columbia* on STS-1, the first mission of the Shuttle era, is now only months away. "Reportorial" artist Tracy Sugarman records the scene outside the VAB.

"Preparations of the Space Shuttle *Columbia*"
During the countdown for the first launch of the Shuttle and of *Columbia,* activities on the launch pad continue day and night. During the built-in holds in the countdown the technicians are able to make up lost time. This painting is by Mark McMahon.

"1:07:22 and Counting"
During the early morning of April 12, 1981, the tension at the Kennedy Space Center is palpable, nowhere more so than at the press site. Artist Arthur Shilstone is among the spectators looking toward the launch pad, brightly lit by xenon lights. He records the scene as the illuminated countdown clock shows 1 hour, 7 minutes, and 22 seconds before lift-off.

"Launch of the *Columbia* — STS-1"
A fraction of a second after 7:00 A.M. local time on April 12, 1981, *Columbia* lifts off the launch pad for the first time. The gleaming, pristine orbiter, crewed by John Young and Robert Crippen, leaves behind a pad enveloped by fire and smoke. Booster and external tank separations take place on schedule, and within 15 minutes, *Columbia* is climbing into orbit. Chet Jezierski painted this fish-eye view of the launch.

"The Right stuff, on Final" Moments away from touchdown on April 14, 1981, *Columbia* drops steeply down to the runway at Edwards Air Force Base in California, accompanied by a chase plane, in this painting by William Phillips. Touchdown occurs at 10:21 local time. The mission has lasted 54 hours and 20 minutes.

◄
"Lift-off of the *Columbia* — STS-2"
After its triumphant first flight in spring 1981, *Columbia* returns by jumbo to the Kennedy Space Center to be prepared for a second launch in the fall. A launch attempt on November 4 is aborted just 31 seconds before lift-off. And not until November 12 does *Columbia* soar off the pad for an unprecedented second journey into space. Paul· Salmon's painting records this historic lift-off.

►
Radar imaging
On its second mission (STS-2), *Columbia* carries its first payload, a package of scientific instruments prepared by NASA's Office of Space and Terrestrial Applications. One of the instruments is the Shuttle imaging radar (SIR), which uses radar to build up a picture of surface topography. This false-color SIR image shows part of the Hammersley Mountains in Western Australia. The colors show the nature of the terrain. Red, for example, shows rugged areas; green shows deserts; blue shows flat areas, such as dry stream beds.

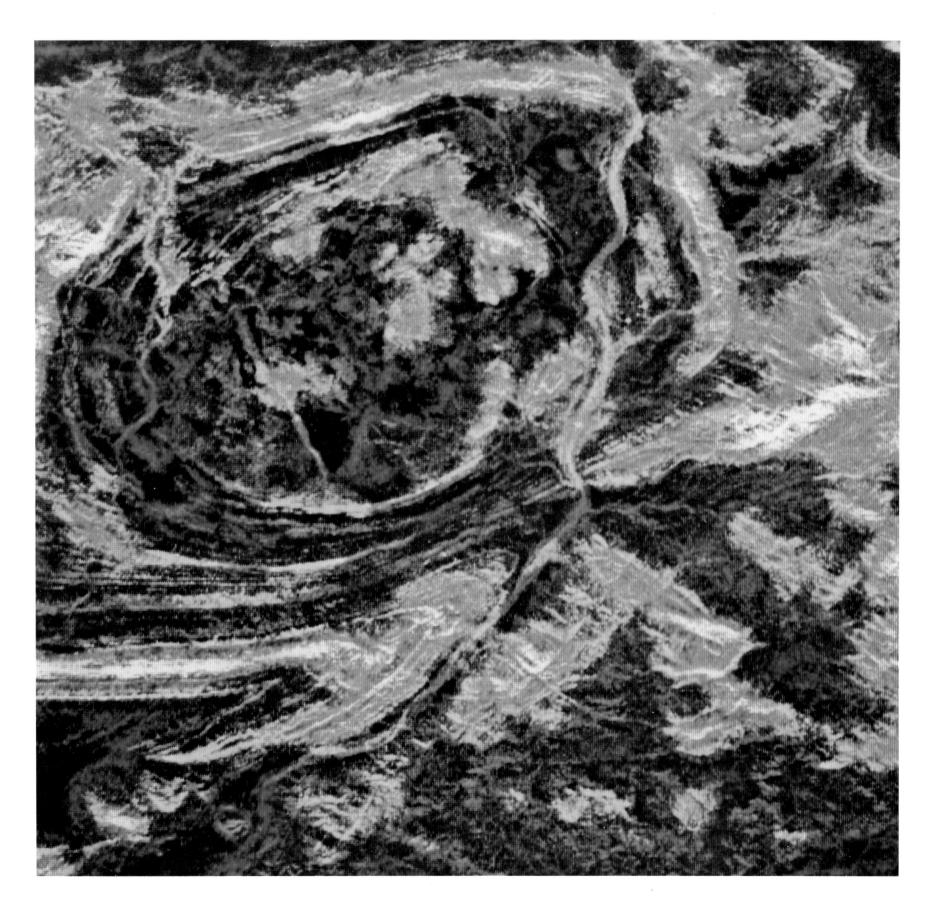

"Shuttle Night Power"
At 2:32 A.M. on August 30, 1983, the skies around Cape Canaveral turn bright as day as *Challenger* thunders away from the launch pad to make the Shuttle's first night flight (STS-8). Photography from the press site is a hit-and-miss affair because there is no benchmark for film exposure. In this painting, however, artist Tom Newsom moves in close, right into the glaring lights of the booster exhaust.

▲
Night landing
For good measure, on mission STS-8, *Challenger* also lands at night. This photograph shows the orbiter on the point of touching down at Edwards Air Force Base at 40 minutes past midnight on September 5, 1983. The night launching and landing demonstrate the 24-hour operating capability of the Shuttle system.

SHUTTLE MISSIONS

◀

"The Landing — *Columbia* 3"
At the end of its third mission
(STS-3) in March, 1982,
Columbia is unable to land as
intended at Edwards Air Force
Base because of flooding. It is
switched to the back-up
landing site at White Sands
Missile Range, New Mexico.
Strong winds at the site delay
landing until March 30, a day
later than scheduled. In this
painting of the landing, Jack
Perlmutter depicts a highly
imaginative and symbolic
landscape.

▶

"Florida Homecoming"
At 7:15 A.M. on February 11,
1984, *Challenger* glides in
over the misty subtropical
landscape to make the first
Shuttle landing at the Kennedy
Space Center. Artist Kent
Sullivan's painting captures the
occasion, the first time that a
space vehicle has returned to
its launch site. The launch pad
Challenger left eight days
before lies just 4 miles
(6.5 kilometers) away.

In-orbit operations

Satellite launching is the prime purpose of the Shuttle, and it is able to accommodate several satellites at once in its cavernous payload bay. Payloads of up to 25 tons can be carried into low orbit. Some are launched by the Shuttle's "crane", the 50-foot (15-meter) long arm of the remote manipulator system. Built by Spar Aerospace in Canada, this piece of equipment is called the Canadarm.

▼
The versatile MMU

On February 4, 1984 (mission 41-B), Bruce McCandless makes the first untethered spacewalk in the MMU, the manned maneuvering unit. Firing thrusters powered by compressed nitrogen gas, he ventures more than 300 feet (90 meters) away from *Challenger*. On later missions the MMU proves its worth in spectacular satellite-recovery operations.

▶
Capturing Solar Max

George Nelson flies the MMU on mission 41-C (April, 1984) in an attempt to capture the ailing satellite Solar Max. This artwork depicts him closing with the satellite, with a T-pad grappling device strapped to his chest. It is designed to mate with a device on Solar Max. In the event, dangerous gyrations of the satellite foil his efforts. But later Solar Max is captured by *Challenger*'s RMS arm.

Back in business

The tragedy of *Challenger*, which fell in pieces out of the Florida skies on January 28, 1986, halts Shuttle activities for 32 months. On September 29, 1988, however, the Shuttle returns to space as *Discovery* punches its way through the clouds. Wendell Minor was among the six artists present to record the Shuttle's rebirth. He is seen here with three of his works. The large one, entitled "Green for Go — America Returns to Space", reflects his wonder at the coexistence of man-made technology and nature in a wildlife preserve.

▶▶

Per ardua ad astra

The crew patches of Shuttle missions since the *Challenger* tragedy have often incorporated seven stars to commemorate the seven astronauts who perished. They appear, for example, in this crew patch for STS-27, the second post-*Challenger* flight by *Atlantis*, which lifted off on December 2, 1988. The rainbow in the design symbolizes the triumphant revival of the nation's manned space program.

◄

"Hot shot"
The pieces of art on these pages illustrate the way in which artists are, in art critic Frank Getlein's words, "pushed by the material itself into taking an abstract, symbolic view of the space flights". Robert Rauschenberg created this lithograph, inspired by the first flight of the Space Shuttle, to "share and express the artist's beliefs in the spiritual and psychical improvement of life and mind through curiosity".

►

"Metamorphosis"
Artist Ingrid Leeds was inspired by the first operational flight of *Columbia* (STS-5) to execute this silkscreen diptych, which alludes to mankind's watery past and to the unfolding of new perspectives.

6

From *Skylab* to *Freedom*

"*Freedom*'s permanency and its mix of human aptitudes and technologies will add an extraordinary dimension to the achievement of long-term space goals."

Leonard David in Space Station Freedom, *NASA Office of Space Station.*

Redundant hardware left over from cancelled Apollo mission provided the unlikely starting point, in the early 1970s, for one of the most influential projects in space history. This was the *Skylab* project. *Skylab* was an experimental space station as roomy as a two-story house. Over a nine-month period in 1973/74, it was home to three teams of astronauts who lived on board for a record-breaking 28, 59, and 84 days respectively.

The $2.5 billion project was almost written off during the launch phase, and only do-it-yourself repairs by the astronauts in orbit saved the day. This provided the first demonstration of the advantage of the human touch when things go wrong in space.

The *Skylab* astronauts carried out prolonged investigations in many branches of science and engineering, particularly in the study of the Sun. They also used themselves as guinea pigs to monitor the effects of the space environment on the human body over a long period. Above all, they demonstrated that, with the correct diet and proper exercise, human beings can live and work in space for months at a time without suffering permanent ill-effects.

NASA hoped that *Skylab* would survive into the Shuttle era so that it could be refurbished and used again. But it did not. So far, space science has played a relatively minor role on Shuttle missions since there is only limited space on board the orbiter for experimentation. Only on the occasional Spacelab flight, of which there were just four in the pre-*Challenger* era, has prolonged scientific investigation take place. Spacelab is a fully equipped European-designed laboratory that fits inside the Shuttle orbiter's payload bay. It remains there throughout the mission.

Modules very much like Spacelab will provide the nucleus of NASA's next mammoth project, the *Freedom* space station. This next logical step in the exploitation of space will see Spacelab-type modules being ferried into orbit by the Shuttle, perhaps as early as 1995. Once in space, they will be linked together to form a structure that will eventually measure over 500 feet (150 meters) across.

When fully operational at the turn of the century, *Freedom* will provide the springboard for other ambitious ventures, such as the construction of solar power satellites and missions to the Moon and Mars.

Skylab in orbit

The second crew to visit *Skylab* snap this picture of the space station before returning to Earth on September 25, 1973. Early in the mission they erected a second sunshade over the one the first *Skylab* team put in place over the area ripped away during lift-off.

Weightless antics

Skylab astronaut Gerald Carr takes time out to perform some weightless gymnastics on the final *Skylab* mission. He is pictured in the cavernous forward compartment above the living quarters.

Laboratory in the sky

Over the years NASA has had more cause than most to lament the fickleness of public and government opinion. A classic example was the Apollo project. After sharing in the euphoria about beating the Soviets to the Moon, people became bored and mean. Where were the benefits, they asked, in making more Moon-landings? So budgets were slashed and the last three landing missions were cancelled.

This left NASA with surplus hardware, of which it was not slow to take advantage. It used an empty S-IVB rocket, the upper stage of the Saturn V, as the nucleus of an experimental space station. It also had available a Saturn V rocket to launch the station into orbit, and three spare Apollo CSM (command and service modules) to ferry up three crews. In 1970 the project was given the name *Skylab*.

The *Skylab* cluster

The S-IVB rocket made up *Skylab's* orbital workshop (OWS). The living accommodation occupied the tank that should have contained liquid hydrogen fuel. The main part of the OWS carried the crew quarters, comprising a wardroom, sleep compartment, experiment compartment, and waste-management compartment ("waste-management" is a NASA euphemism for "toilet").

What would have been the liquid oxygen tank beneath the living quarters was used for storing trash. The upper story of the OWS was a storage area, which also doubled as an exercise yard. The whole OWS was pressurized with an Earth-type atmosphere (nitrogen/oxygen) but at only one-third of normal sea-level pressure.

A hatch at the upper end of the OWS led to an airlock module (AM), which could be separately depressurized when the astronauts went spacewalking. The AM was joined to the multiple docking adapter (MDA), which carried two docking ports into which the Apollo ferry craft could dock. With the ferry craft docked, the whole structure measured 120 feet (36 meters) from end to end and weighed nearly 100 tons.

Power for *Skylab* was provided by an X-shaped, windmill-like solar array mounted on the MDA and a flag-like solar array on the OWS. There should have been a second, which brings us to the *Skylab* launch.

Making do and mending

Skylab was sent into orbit on a Saturn V rocket on May 14, 1973. It looked like a perfect lift-off, like those of all the other Saturn Vs. But telemetry coming back from the space station when it reached its 270-mile (435-kilometer) orbit revealed that all was far from perfect. It appeared that during lift-off one of the solar arrays on the OWS had been ripped off, along with a large area of meteoroid and heat shielding. The other OWS panel appeared to be jammed. As a result the station was starved of power and starting to overheat.

The planned launch next day of the first *Skylab* crew was aborted, while engineers worked out what exactly had gone wrong and how to put it right. The decision was taken to send up the crew and see if *they* could somehow fix the problem, assuming it was fixable.

Accordingly, on May 25, Charles Conrad, Joseph Kerwin, and Paul Weitz were launched on a make-do-and-mend mission. They took with them a sunshade to cover the damaged area, wire cutters, and other tools. After failing to fix anything from the outside, they decided to dock and tackle the problem from the inside. Eventually, in 122°F (50°C) heat, they managed to fix the sunshade in place, working through the open hatch of the AM.

Keeping in shape

The temperature inside *Skylab* soon began to fall, and when it had reached a tolerable level the astronauts began their scheduled workload. The lack of power was a nagging problem, which the astronauts remedied on June 7 by freeing the jammed OWS panel. Apart from a few minor malfunctions, *Skylab* was in good shape when it was mothballed after 28 days and the crew returned to Earth.

Doctors pronounced that the crew were in good shape too, which augered well for long-term living in space. The astronauts had kept up a rigorous exercise regime in orbit, working out on a bicycle ergonometer and running round and round the storage compartment in the OWS.

The second *Skylab* crew of Alan Bean, Owen Garriott, and Jack Lousma took up residence in *Skylab* on July 28. They were immediately struck with severe space sickness, something that had not affected the first crew. Fortunately, the condition lasted for only a few days and soon everyone was hard at work. They overshot their planned stay in orbit by three days, returning to Earth on September 25.

After 59 days in space — more than doubling the space duration record — they were still in remarkably good shape. There was some weakening of heart muscle and some loss of calcium from bones, and their leg muscles had lost bulk and become flabby. But these were temporary effects, and within a week everything was more or less back to normal. The third crew, Gerald Carr, Edward Gibson, and William Pogue, visited *Skylab* for no less than 84 days, from November 16, 1973, through February 4, 1974. Astonishingly, they were the fittest crew of all when they returned!

Altogether during the *Skylab* missions, the crews flew more than 2,500 times around the Earth and traveled more than 70 million miles (115 million kilometers). The 84-day duration of the final *Skylab* crew is still the longest any American has remained in space. Only when space station *Freedom* becomes operational will this record be beaten. Soviet cosmonauts, however, have bettered this many times over. In 1987 Anatoli Levchenko and Musakhi Manarov spent exactly one year in orbit in the Soviet space station *Mir*.

Laboratory in space

Some scientific experimentation takes place on the mid-deck of the Shuttle orbiter in the crew quarters of the spacecraft. Usually experimental modules are fitted into one of the storage lockers. But this area, which serves as lounge, dining room, kitchen, bathroom, and sleeping quarters for up to seven people, is hardly ideal for painstaking laboratory work!

To introduce a higher level of science into the Shuttle program, the Spacelab project was initiated. Spacelab, as the name implies, is a fully equipped laboratory in which scientists, doctors, and engineers can carry out experiments in space in many scientific disciplines. Spacelab hardware, which comprises a variety of modules of different size and function, was designed and manufactured in West Germany and funded by ESA.

ESA, the European Space Agency, is the European equivalent of NASA, which coordinates space activities in Europe. It has 13 member states — Austria, Belgium, Denmark, France, West Germany, Ireland, Italy, the Netherlands, Norway, Spain, Sweden, Switzerland, and the United Kingdom.

Spacelab's maiden flight

Spacelab is made up of two basic parts, a pressurized laboratory module and an unpressurized pallet, which carries instruments that need to be exposed to the space environment. Scientists and engineers carry out their experiments in the laboratory module, which usually consists of two standard segments, together measuring some 30 feet (7 meters) long and 13 feet (4 meters) across. The module is fitted with state-of-the-art equipment and powerful computing facilities. Experimental hardware is stowed in standard-size racks around the sides of the module. In the roof there is a window of optically pure glass to permit high-resolution photography. There is also an airlock in which scientists can expose materials to the vacuum of space.

The first flight of the space laboratory, *Spacelab 1*, on Shuttle mission STS-9, took place on November 28, 1983. It was carried into orbit by *Columbia*, making its first flight for a year. As on all Spacelab flights, the module remained inside the orbiter's payload bay.

Spacelab 1 was a record-breaking flight for many reasons. *Columbia* carried a crew of six, the most that had ever flown into space together. The commander of *Columbia* was 53-year old John Young, a veteran of the Gemini and Apollo eras, who was making a record sixth flight into space. The mission was also the first to fly a new category of "occasional astronaut", the payload specialists. One of these was the West German Ulf Merbold, who became the first non-American to fly on a U.S. spacecraft.

The *Spacelab 1* crew worked two 12-hour shifts around the clock, resting in *Columbia*'s crew cabin between times. They carried out more than 70 experiments in biology, medicine, materials science, Earth observation, and astronomy. The success of the mission was outstanding. When *Columbia* returned to Earth on December 8, it set yet another record — for the longest Shuttle flight to date, 10 days 7 hours.

The European connection
The West German physicist Ulf Merbold at work during the inaugural flight of the European-built Spacelab in December, 1983. Merbold and the other five crew members on this *Spacelab 1* mission (STS-9) work in shifts around the clock for ten days.

Eight's a crowd
On the final Spacelab mission of 1985, *Spacelab D1*, *Challenger* provides cramped quarters for the largest number of astronauts ever to go into space together. The eight-member crew poses here for the automatic camera. They are, from left to right: at top, Henry Hartsfield, Bonny Dunbar, James Buchli, and Reinhard Furrer; below, Ernst Messerschmid, Wubbo Ockels, Steven Nagel, and Guion Bluford.

The flying zoo

Orbiter *Challenger* took Spacelab into orbit on its second flight, actually designated *Spacelab 3*. The 7-day mission, Shuttle flight 51-B, began on April 29, 1985. It was the first of three Spacelab missions that year. Irreverently nicknamed "the geriatric flight", it featured three astronauts over 50 years old. William Thornton, at 56, was the oldest man ever to have made a space flight, proving that on the Shuttle at any rate advancing age is no barrier to space travel.

The seven human crew were accompanied into orbit by a pair of monkeys and two dozen rats. These were used as test subjects for studies in space adaptation syndrome, or space sickness. It was the first time that humans and animals had been placed in space together, and from the animal waste-management viewpoint it was not a success. The presence of weightless monkey wastes floating about in the air did not find favor with the human crew!

One unexpected bonus during the mission was that the crew were treated to a spectacular display of the aurora australis, or Southern Lights. Auroras, caused by the interaction of charged particles with atoms in the Earth's upper atmosphere, take place mainly in polar regions. This particular one was observed as a rippling curtain of shimmering light stretching all the way from the Antarctic to Australia.

Abort to orbit

Challenger also carried the next two Spacelabs into orbit. On *Spacelab 2*, Shuttle mission 51-F, it was the turn of Karl Heinze to beat the space age record: he was aged 58.

At the first launch attempt on July 12, the mission was aborted just three seconds before lift-off when one of the main engines developed a fault, and all three shut down. And nearly five minutes after lift-off on July 29, the computers shut down one of the main engines when a temperature sensor registered critically high values. This left the crew in an abort-to-orbit situation, which would place them in a safe, but lower orbit than planned.

Then the sensor on one of the two remaining engines also began registering high values. This could have resulted in the engine shutting down and precipitating the first really critical in-flight emergency for a Shuttle crew. Engineers decided that the sensor must be faulty and told the crew to overrule the computer which would have shut the engine down. Fortunately, they were right, and *Challenger* reached orbit safely.

After all this excitement, the mission proceeded as planned. It was an instruments-only Spacelab, just using unpressurized pallets in the payload bay. The 13 major experiments, mainly in astrophysics and astronomy, yielded, said NASA, "sensational results".

Deutschland über Erde

Spacelab D1, the final flight of Spacelab in 1985, was unusual in that it was dedicated to West Germany, which footed the $175 million bill. It was under the control of West Germany's space operations center at Oberpfaffenhofen, near Munich.

Challenger carried *Spacelab D1*, on Shuttle mission 61-A, into orbit on October 30. It was the ninth and last time that this orbiter would venture into space. One of the most interesting of the 76 experiments attempted on this mission involved the use of the so-called space sled. This was a contraption on which astronauts were accelerated along a track. They wore elaborate headgear that exposed eyes and ears to different stimuli. The experiment was devised to acquire further information into the causes of space sickness, the problem that plagues one in two astronauts for the first few days in orbit.

Writing about the mission, *Flight International*'s space correspondent Tim Furness commented: "...it was an important milestone on the long road to the space station, with the Europeans illustrating how well they could collaborate with NASA ... a dress rehearsal for permanent space operations."

A space station called *Freedom*

With the Shuttle program back on course, NASA's upcoming priority is the construction in orbit of a permanent space station. President Ronald Reagan sanctioned the project in his State of the Union message in January, 1984. "We can follow our dreams to distant stars, living and working in space for peaceful, economic, and scientific gain," he said. "Tonight I am directing NASA to develop a permanently manned space station and to do it within a decade." Four years later, he announced that the space station would be called *Freedom*.

NASA and America will not go it alone in developing *Freedom*. From the beginning, President Reagan invited other nations to join this exciting venture which will shape the free world's future in space. Canada, Europe, and Japan have accepted the invitation and are making significant contributions to the program. They are providing major items of hardware, although the actual construction work will be left to NASA.

Unlike the space stations that have gone before it — the American *Skylab* and the Soviet *Salyut*s and *Mir* — *Freedom* is too big to be lofted into orbit as a whole. It is anticipated that at least 20 Shuttle flights, starting in the mid-1990s, will be needed to ferry up the components of the station.

However, this number may be slashed if an alternative scheme is employed, using an unmanned cargo version of the Shuttle known as Shuttle C. This scheme uses an unmanned orbiter — the expendable payload carrier — fitted with Shuttle main engines near their scrap date. Like the manned version, Shuttle C draws its propellants from an external tank and blasts off with the aid of twin recoverable solid rocket boosters.

Under construction

The first stage in the construction of *Freedom* will be to put together the long truss assembly that will form the backbone, or keel, of the space station. Astronaut-engineers will do this using a method similar to that demonstrated by spacewalking astronauts Jerry Ross and Sherwood Spring in November, 1985. On Shuttle mission 61-B they assembled long trussed beams from sections that snap-fitted together. They demonstrated two techniques, one called ACCESS (assembly concept for construction of erectable space structures) and the other EASE (experimental assembly of structures in extravehicular activity).

Hopefully, the construction of the keel will be accomplished on the first Shuttle flight. A pair of solar panels and associated electronic modules will also be fitted at the same time. Later flights will ferry up four pressurized modules that will be interconnected to form

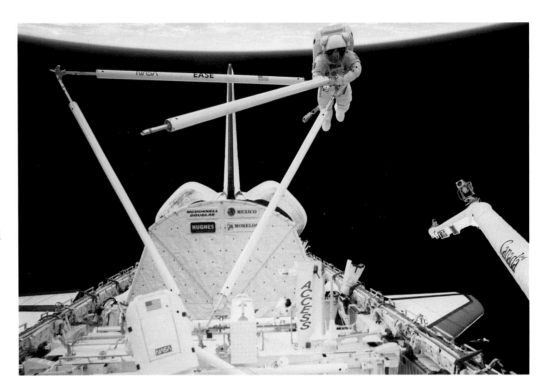

the nucleus of the space station.

The first two modules to be put in place will be the U.S. laboratory module and the habitation module, both supplied by NASA. Later, the European Space Agency (ESA) and Japan will supply additional laboratory modules. The habitation module will provide accommodation for the station crew, who will probably number six. There will be a crew change every three months to start with, but this may be increased to six months or even longer for space-medicine studies.

The construction of *Freedom* will be greatly aided by Canada's contribution to the station, a mobile servicing system (MSS). This will feature a robot remote manipulator arm mounted on a module that can move along the keel. It will be a development of Canadarm, the remote manipulator system arm used to handle payloads in the Shuttle orbiter's payload bay.

By the turn of the century *Freedom* should be fully operational. It will not only be the focus of most zero-gravity medical, scientific, and engineering studies. It will also take over the role of retrieving and servicing payloads in orbit, for example, the Hubble Space Telescope. For this expanded role, a space tug called the orbital maneuvering vehicle (OMV) will be developed. This will be able to retrieve satellites under remote control. Later, an orbital transfer vehicle (OTV) will be built as an orbital runabout for astronauts, who will use it to visit the free-flying platforms that will form part of the space station set-up.

EASE and ACCESS
Jerry Ross and Sherwood Spring practice techniques for building extended space structures on mission 61-B in November, 1985. The techniques they use, EASE and ACCESS, enable them to build quickly and accurately, despite their cumbersome spacesuits.

Skylab II
This was the crew emblem for the second manned *Skylab* mission. Its dominant feature is a figure drawn by Renaissance artist and inventor Leonardo da Vinci to illustrate the proportions of the human body. Its use here suggests the many physiological studies to be conducted on the mission in the zero-g environment. The two hemispheres behind the figure represent the other two main areas of research — studies of the Sun and of the Earth's resources.

◄◄
First to *Skylab*
Because of the near-disaster at the launch of *Skylab*, the first crew — Charles Conrad, Joseph Kerwin, and Paul Weitz — are not ready to lift off until May 25, 1973. Here, artist Ron Brown portrays them suited up and ready to enter the Apollo capsule on top of the Saturn IB rocket. This is shown at right, perched on its launch tower.

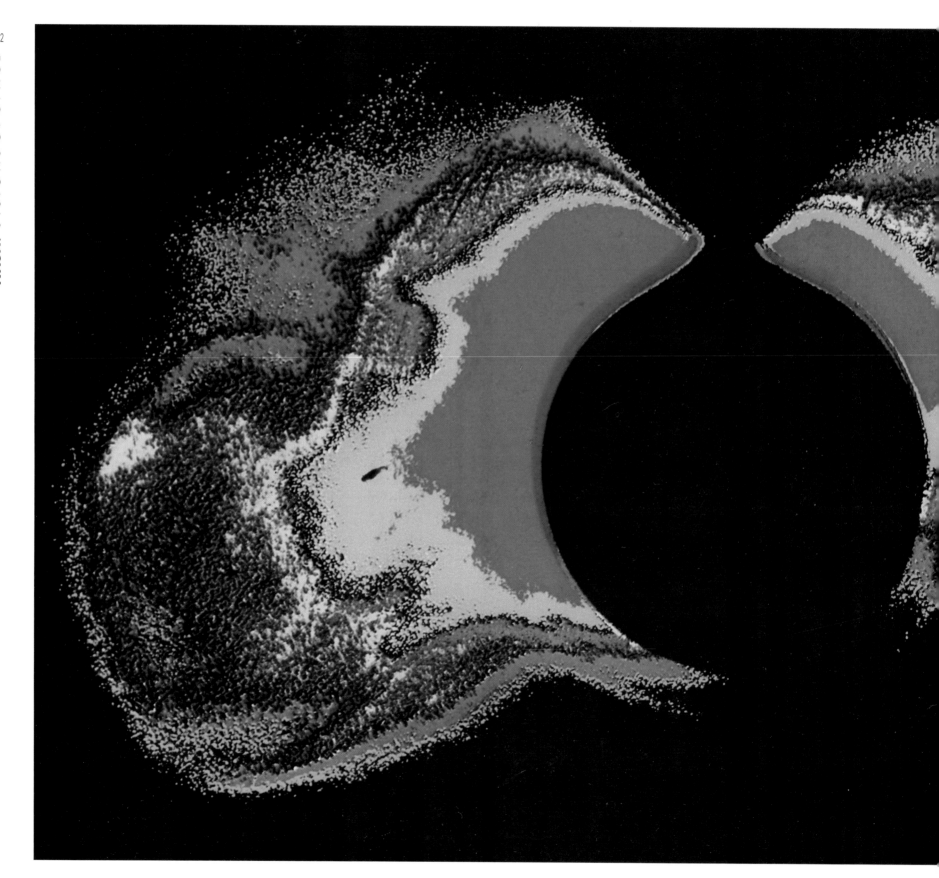

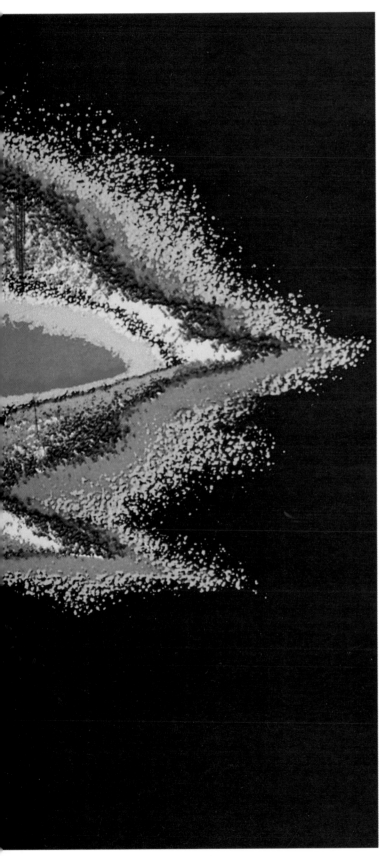

► **Tongues of fire**
An enormous prominence, a tongue-like fountain of flaming gas, leaps high above the Sun's surface. This gives rise to more widespread disturbances over distances of millions of miles.

◄ **The Sun's crown**
The most spectacular scientific results from the *Skylab* missions came from the astronauts' detailed studies of the Sun. This picture shows, in false colors, different levels of brightness in the Sun's outer atmosphere, or corona (crown). The corona has been made visible by producing an artificial eclipse in an instrument called a coronagraph.

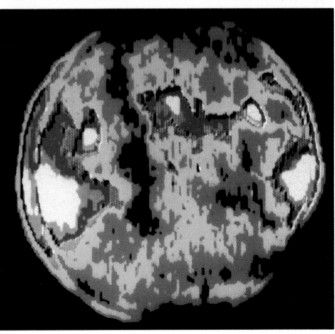

► **Holes in the corona**
One of *Skylab*'s most important discoveries was of "holes" in the corona. They appear as dark areas in pictures taken at ultraviolet wavelengths, as here, and in X-rays. They are important because they appear to be regions through which charged particles can escape from the Sun and cause magnetic storms on Earth.

Made in Europe
The emblem of the first mission of Spacelab, carried into space by orbiter *Columbia* on the ninth Shuttle flight, reflects the international parentage of the space laboratory. It was designed and built in Europe, under the auspices of the European Space Agency (ESA). At the time of the mission (November, 1983), ESA was composed of ten member countries, whose flags appear on the emblem together with the Stars and Stripes.

▼
The laboratory module
This artwork shows the laboratory module in more detail. The instruments and experimental equipment are mounted in standard-sized racks around the walls. There is an airlock for experiments in the roof, together with a window with optically clear glass for photography. This version of Spacelab, called the long module, is about 30 feet (7 meters) long and 13 feet (4 meters) in diameter.

▶
Spacelab in orbit
This cutaway artwork shows how Spacelab fits inside the Shuttle orbiter. It remains there throughout the mission. Spacelab is made up of a pressurized laboratory, where the crew work, and a pair of unpressurized pallets carrying instruments. Access to the laboratory is via a tunnel from the mid-deck of the orbiter.

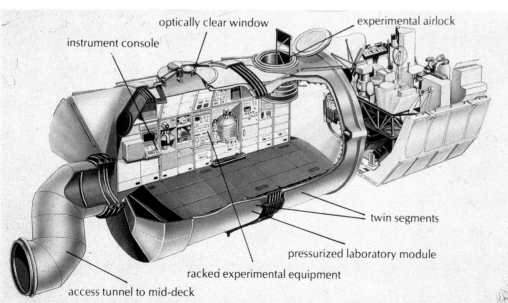

optically clear window

instrument console

experimental airlock

twin segments

pressurized laboratory module

racked experimental equipment

access tunnel to mid-deck

The modular approach
A close-up of the central complex of a space station, as visualized in the mid-1980s by Boeing Aerospace, now one of the prime contractors for the construction of space station *Freedom*. It shows four cylindrical modules, which provide laboratory facilities and living quarters for the crew. At top right a remote-controlled manipulator arm is holding a satellite that needs servicing. The docked Shuttle orbiter gives an idea of the scale of the station.

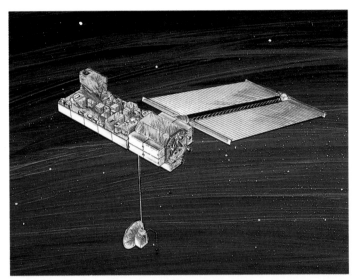

Free-flyer
Free-flying spacecraft also feature in the space station concept. This U.S. polar-orbiting platform is designed to carry a variety of remote-sensing instruments as part of NASA's comprehensive Earth observation system (EOS) program. It will be serviced from the space station.

▼
Orbital transfer
Boeing also visualized this kind of reusable rocket stage as a "space tug" for maneuvering spacecraft and transferring them to high orbit. Later versions would be manned.

Steady as she goes
A Space Shuttle orbiter prepares to dock with space station *Freedom* at the turn of the century, in this painting by Alan Chinchar. It is docking at a port on one of the resources nodes that link the U.S. laboratory and habitation modules. The resources nodes at the other end of the modules mate with the European- and Japanese-built laboratories. In the foreground is one of the eight solar arrays, each measuring 95 by 31 feet (29 by 9.5 meters). Together they produce 75 kilowatts of electricity. They are located at either end of the 500-foot (150-meter) long transverse boom, which forms the backbone of the station.

The American connexion
This cutaway shows in close-up the U.S. science laboratory module (top), and the habitation module, which will be used by all *Freedom* crew members. Each module is nearly 45 feet (13.6 meters) long and about 15 feet (4.5 meters) across. The interconnecting resources nodes carry docking ports. The docked logistics module (upper left) will be ferried into orbit by the Shuttle, carrying fuel and perishables to replenish space station supplies.

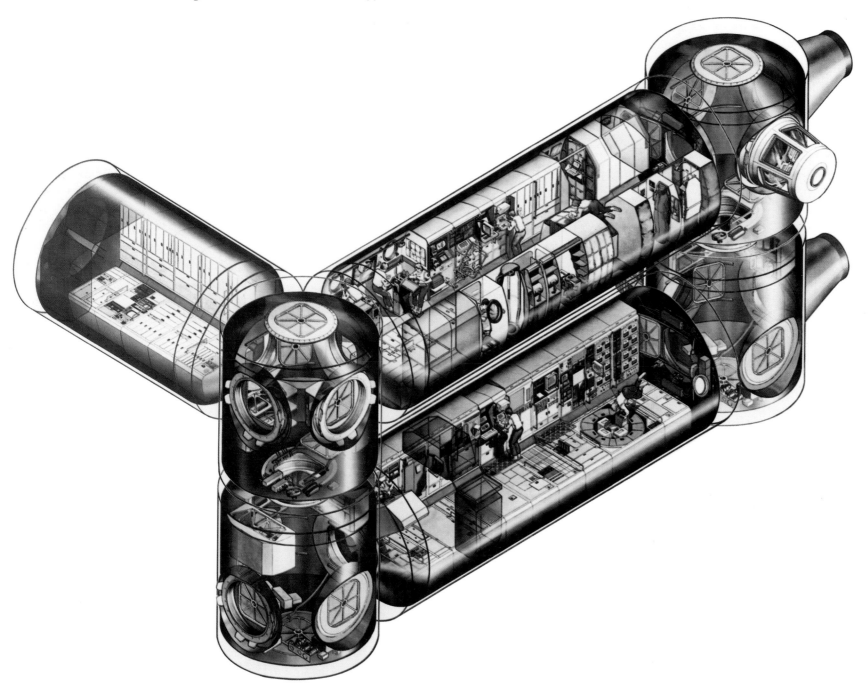

Space spidermen
In the first decade of next century work is well underway on the biggest structure yet attempted in space, a solar power station. There is nothing fundamentally difficult about building such a station in orbit, except the scale. To capture enough energy, solar collectors must have gargantuan dimensions, measured in terms of miles rather than feet. Much of the work will be carried out by remote-controlled robot manipulators.

Space city
The year is now 2090. An astronaut is carrying out maintenance work on a rotating wheel-shaped habitat floating in deep space. Earth is 240,000 miles (385,000 kilometers) away. Reflected in the gold-tinted visor of the astronaut's space helmet is another identical habitat close by. Each measures nearly 1 mile (1.6 kilometers) across and has a population of 10,000 people. Some of the younger ones have never been to Earth.

17 To other worlds

"Man will not remain on Earth forever. In his pursuit of light and space he will at first timidly probe beyond the atmosphere, then conquer all of circumsolar space."

Konstantin Tsiolkovsky, the father of astronautics.

Racing through space at speeds of up to 108,000 miles per hour (180,000 km/h), robot explorers from Earth have taken us to seven of the other eight planets that make up the solar system. They are the spacecraft we call probes, which NASA has been lofting into space since 1962. Among the planets only Pluto, a deep-frozen world much smaller than our Moon, has escaped attention.

Through the TV eyes of these miracles of electronic technology — the Mariners, Pioneers, Vikings, and Voyagers — what sights we have seen. Sun-baked lunar-like landscapes on Mercury. Swirling clouds of sulfuric acid on Vénus. River channels and great canyons on the red planet Mars. Raging volcanoes on Jupiter's pizza-like moon Io. Tiny moons shepherding the particles in Saturn's rings. The most baffling geology in the solar system on Uranus's moon Miranda. And erupting geysers on Neptune's moon Triton.

Most of the probes NASA has launched are destined to circle forever in interplanetary space, held in the grip of the Sun's powerful gravity. But four probes — two Pioneers and two Voyagers — are traveling fast enough to escape into interstellar space. In the far-distant future they could come under the influence of other suns, even other planets. In case they do, they carry messages from Earth for any intelligent extraterrestrials who might happen to find them. On the Pioneers the message takes the form of a pictorial plaque. On the Voyagers it is more elaborate, appearing on a record that provides sounds and sights of Earth.

These terrestrial ambassadors will finally break free of solar gravity early next century. By then humankind will be taking further strides to establish a permanent foothold on other worlds. Space station *Freedom* will be operational. This will support the construction of a permanent base on the Moon. It will also provide springboard for the next "giant leap" into the solar system — manned exploration of the red planet Mars.

Such a mission could be attempted in the second decade of next century, but it would require a massive national commitment comparable with that required to plant the first human footprints on the Moon. Perhaps all the world's space powers will collaborate to mount an international expedition, as would befit such a milestone in the history of the human race.

Escape from Earth

Launching a satellite into orbit around the Earth is difficult enough — it must be boosted to orbital velocity, a speed of some 17,500 miles per hour (28,000 km/h). But in orbit it is still bound by Earth's gravity. To escape completely from this gravitational noose, a spacecraft must be propelled away from Earth at escape velocity, a speed of no less than 25,000 miles per hour (40,000 km/h).

With characteristic optimism, NASA began trying to beat the gravity barrier in August 1958, only seven months after the launch of the first U.S. satellite. The target, inevitably, was our neighbor in space, the Moon. The booster rockets of the day and ancillary technologies were just not up to the task of boosting a spacecraft to escape velocity, let alone of propelling it to an accurate rendezvous with the Moon. Only after four failures did the *Pioneer 4* probe, in March, 1959, make a respectable flyby of the Moon at a distance of about 37,500 miles (60,000 kilometers). Two months earlier the Soviet probe *Luna 1* had got ten times closer. NASA had no complete success in targeting the Moon until the probe *Ranger 7* crash-landed there in July, 1964.

However, encouraged by its partial success with *Pioneer 4*, NASA launched the next probe, *Pioneer 5* in March, 1960, to explore deep space. And highly successful it proved, returning data about conditions in the hitherto unexplored interplanetary medium from a distance of some 22 million miles (36 million kilometers).

Mariners to Venus

Pushing forward with increasing confidence, NASA next set its sights on Venus, the planet nearest to the Earth, which approaches to within 26 million miles (42 million kilometers). This is hardly close by terrestrial standards, but in cosmic terms it is a stone's throw.

The first attempt to target Venus, with the *Mariner 1* probe, foundered when the probe failed to reach Earth orbit. But *Mariner 2* made a successful encounter with the planet on December 14, 1962, approaching to within 22,000 miles (35,000 kilometers). Its heat-sensing instruments made the astonishing discovery that Venus has a very dense atmosphere and that the temperature on the surface is at least 800° F (425°C) — hot enough to melt tin and lead!

Subsequent probes sent to Venus by both the U.S. and the U.S.S.R. have confirmed these results. Venus is a hellish world on which the surface temperature climbs as high as 900°F (480°C). The atmospheric pressure at the surface is a crushing 90 or 100 times that on Earth. This is due to the main atmospheric gas, heavy carbon dioxide. This is the infamous gas that on Earth is the

Pockmarked Mercury
The surface of Mercury from *Mariner 10*. It is pockmarked with craters, formed by meteorite impact in the early days of the solar system.

major cause of the greenhouse effect, which is slowly but inexorably increasing the temperature of our planet. Venus has suffered from a galloping greenhouse effect, which is unlikely ever to be reversed. If drastic steps are not taken to curb the production of greenhouse gases on Earth — a near twin of Venus in size — our planet may in the long term suffer a similar fate.

Mariners to Mars

The Russians too had attempted to reach both Venus and our next nearest planetary neighbor, Mars. Their first near-success came with *Mars 1*, which passed within 120,000 miles (193,000 kilometers) of the Red Planet in June, 1963. Unfortunately, communications with it had failed three months earlier.

Again it was a U.S. probe that eventually made the breakthrough, *Mariner 4*. In mid-July, 1964, it sent back more than 20 pictures of Mars over a distance of some 140 million miles (220 million kilometers). Considering the technology of the day, this was a fantastic feat of communications. The pictures showed craters and evidence of some haze in the atmosphere. This atmosphere proved to be a scant one-hundredth as dense as Earth's and made up mainly of carbon dioxide. This came as a sore disappointment to those who still harbored the hope that there could be life on Mars.

More detailed investigations in 1969 by flyby probes *Mariner 6* and *Mariner 7* confirmed that Mars is an inhospitable planet with a variety of interesting terrains. They gave only a hint of the feast planetary scientists would be treated to when *Mariner 9* went into orbit around Mars in November, 1971.

The first *Mariner 9* views were disappointing because Mars was almost completely obscured by a planet-wide dust storm. Then, slowly and tantalizingly, the dust began to settle. Wonder upon wonder began to emerge — four huge extinct volcanoes, one (Olympus Mons) three times as tall as Mount Everest; a Martian Grand Canyon up to 3,000 miles (5,000 kilometers) long; glistening polar caps of water ice; and channels that probably once carried flowing water.

Gravity-assist

NASA planetary scientists, their appetites whetted by success at Venus and Mars, now began to plan assaults on the outer planets. Sending probes to these giant worlds was obviously going to be a problem because of the distances involved. At closest approach, Mars is only about 35 million miles (56 million kilometers) away from Earth. But the next planet out, Jupiter, lies almost three times as far away; Saturn lies twice as far away as Jupiter; Uranus twice as far away as Saturn; and so on.

So journeys of billions of miles would be involved,

which would take years rather than months to accomplish. How spacecraft systems would fare over such periods and what hazards they would face were unknown.

In the event, two things made the task less daunting than it might otherwise have been. In 1965 a graduate student at the California Institute of Technology, Gary Flandro, suggested a means by which journeys to the planets could be considerably shortened. If a probe were directed close to one planet, that planet's gravity would cause it to accelerate. Provided that a suitable aimpoint was chosen, the probe could then be directed at greater speed on to another planet. This was the birth of the gravity-assist principle.

The other event that favored planetary scientists at the time was the very special alignment of the planets in the solar system due to occur in the 1970s and 1980s. It would enable them, using the gravity-assist principle, to send probes to explore the outer planets in a remarkably short time. A similar opportunity would not occur again for another 175 years. The chance was too good to miss.

Accordingly, NASA began drawing up detailed plans in 1969 for an ambitious program to explore all the outer planets. It became known as the Grand Tour. However, the program had to be severely curtailed following space budget cutbacks by Congress. The result was the Mariner-Jupiter-Saturn project, which involved the launch of two identical probes to Jupiter and Saturn. Approval for the project, which was later renamed Voyager, was given in July, 1972.

The pathfinding Pioneers
By this time another space probe, *Pioneer 10*, was four months into a pioneering flight into deep space to blaze a trail for the Voyagers and report back on the two known hazards they would have to face.

One was the asteroid belt, a broad band between the orbits of Mars and Jupiter, in which circle thousands of rocky lumps and miniplanets, up to 620 miles (1,000 kilometers) across. *Pioneer 10* emerged unscathed from the belt, suffering only a few hits by tiny meteoroid particles.

The next testing time for *Pioneer 10* came in December, 1973, when it swooped to within 83,000 miles (131,000 kilometers) of Jupiter's cloud tops. As expected, it encountered heavy bombardment from the Jovian radiation belts, and some of its instruments were put out of action. It nevertheless returned valuable data and pictures with much higher resolution than any that could be obtained through Earth telescopes.

A near-identical craft, *Pioneer 11*, set out for Jupiter a year after *Pioneer 10*. It was launched in a trajectory

The Great Red Spot
Pioneer 11 images the Great Red Spot, a massive storm center, when it encounters the giant planet Jupiter in December, 1974. There is evidence of circulation in the Spot and violent turbulence in the atmosphere around it.

The plain of Chryse
Viking 1 photographs the Martian landscape shortly after touching down on the Red Planet in July 1976. The whole landscape is a rusty reddish-brown color, probably due to the presence of iron oxides. The sky is tinted pink by dust blown up from the surface.

that would take it closer to the giant planet in December, 1974, so that it could receive a high enough gravity boost to direct it towards Saturn for a close encounter in 1979.

The first craft to utilize gravity-assist had been *Mariner 10*, a probe NASA sent to encounter first Venus (February, 1974), then Mercury (March and September, 1974, March, 1975). *Mariner 10* was the first, and as yet the only, probe to investigate Mercury, showing it to be a barren, crater-covered world reminiscent of the Moon.

The design of the Pioneers broke new ground. They had no fewer than 11 instruments, including one called an imaging photopolarimeter. This was able to record images at different wavelengths. When the imaging data were processed at Pioneer mission control at Ames Research Center in California, full-color pictures could be obtained — the first from a space probe.

Since the Pioneers would be traveling hundreds of millions of miles away, they needed a large diameter dish antenna to pick up the radio signals transmitted from Earth and to channel return signals into a narrow beam. Also, because of their increasing distance from the Sun, they had to be fitted with a nuclear power source — radioisotope thermoelectric generators (RTGs) — to provide electricity for their instruments. The power source used by previous probes, solar cells, would not be effective at distances of Jupiter and beyond because the light from the Sun would be too feeble.

The Viking invasion
The late summer of 1975 saw NASA launch another two Viking space probes to Mars. These were designed not only to orbit the planet for an extended period, but also to release a lander that would touch down on the surface. A prime objective for the landers would be to search for Martian life. No one was expecting to find goggle-eyed little green men, but there was just a possibility that there could be some form of lowly life which could tolerate the Martian environment.

The first probe, *Viking 1*, went into orbit around Mars in June, 1976, and was scheduled to put down its lander on the Bicentennial Fourth of July. However, the planned landing site in a region called Chryse proved to be too rough, and a new site had to be chosen, delaying a landing until July 20. *Viking 2*'s landing site also had to be altered because of the terrain. The lander made a successful touchdown in a region called Utopia on September 4.

While the Viking orbiters circled high overhead, mapping almost the entire planet, the landers took photographs, reported on the weather, and began to search for Martian life. The views they sent back of Chryse and Utopia were remarkably similar. They showed a rust-colored landscape strewn with small rocks, with drifts of fine material here and there. The sky had a pinkish hue, presumably due to the presence of fine reddish dust blown up from the surface.

When samples of soil were tipped into the landers' mini-laboratories, some strange reactions seemed to occur. But, despite initial optimism, the results failed to indicate the presence of life. This is not to say that there is no life on Mars. Only when astronauts travel there and search for themselves will we be sure, one way or the other.

Deep-space Voyagers

The Pioneer probes carried out a reconnaissance of the gas giants Jupiter and Saturn, preparing the ground for the two vastly more sophisticated Voyager probes. *Voyager 2* was the first one to be launched, on August 20, 1977. On that day it began a journey the like of which cannot be repeated until the middle of the twenty-second century at the earliest. It has performed the Grand Tour NASA conceived in the 1960s, and performed it in the most remarkable and spectacular manner.

Voyager 2 was launched 16 days before *Voyager 1* but into a slower trajectory. As a result, *Voyager 1* was first to reach Jupiter and afterwards Saturn. By the time it made its closest approach to Jupiter, on March 5, 1979, it had totally transformed our knowledge of that planet and its retinue of moons. It sent back fantastic images of the Jovian system, showing a multicolored banded atmosphere in perpetual turbulence. Nowhere was this more evident than around the Great Red Spot, now seen to be the eye of a gigantic storm that has been raging for centuries.

The star among its four large moons, which had been spotted first by Galileo, was undoubtedly Io, nicknamed the pizza moon because of its bizarre orange, red, and yellow surface. This was found to be caused by sulfur-spewing volcanoes. Nowhere else but on Earth had volcanoes been found before.

Myriads of rings

Voyager 1 looped around Jupiter and was then propelled by gravity-assist toward Saturn. *Voyager 2* encountered Jupiter on July 9, 1979, filled in various gaps in the coverage of the Jovian system, and then departed for Saturn. *Pioneer 11*, meanwhile, now renamed *Pioneer Saturn*, was drawing close to the ringed planet, eventually making a close approach on September 1. It discovered a new moon and two new rings, sending back images at more than 20 times the resolution obtainable from Earth. Disappointingly few features could be detected on the planet's disk.

Then came the close encounters with Saturn by the

The Galilean moons
An artist has added a stellar background to this montage of five *Voyager 1* images taken in March, 1979. It features Jupiter and its four largest moons, first spotted by Galileo in 1609. Io (upper left) lies closest to Jupiter. Then come Europa (center), Ganymede (bottom left), and Callisto.

Voyagers — *Voyager 1* on November 12, 1980, and *Voyager 2* on August 25, 1981. Between them they sent back thousands of extraordinary images that defied explanation at the time and still continue to baffle scientists. Voyager mission scientists were heard to remark that in the abysmal depths of the solar system the bizarre seems to be commonplace.

The three shining rings we see circling Saturn from Earth turn out to be made up of thousands of separate ringlets, formed by rocks and dust traveling at high speed. On either side of some of the rings there are little "shepherd" moons, which somehow keep the ring particles in place. Mysterious dark spokes appear in the rings, seemingly with a mind of their own.

Among Saturn's moons, Titan provided the biggest surprises. It is the only moon in the solar system to have an appreciable atmosphere. This turns out to be mainly nitrogen, with some methane. The speculation is that in Titan's atmosphere methane plays the same role as water vapor does in Earth's atmosphere. So on Titan it could rain liquid methane and there could be methane rivers and lakes.

Twin Uranus and Neptune

After their encounters with Saturn, the two Voyagers went their separate ways. *Voyager 1* began heading out of the solar system, but *Voyager 2* still had a long way to go and two new worlds to conquer — Uranus and Neptune. Both planets were unknown to the ancients because they are too far away to be visible to the naked eye. Even in powerful telescopes no details of their surface can be seen.

Voyager 2 reached Uranus on January 24, 1986. At that time it was nearly 2 billion miles (3 billion kilometers) away from Earth. At such a distance Voyager's radio signals, even though they were traveling at the speed of light, took 2 hours and 45 minutes to reach Earth! Visually the planet was disappointing, appearing quite bland, unlike the banded Jupiter and Saturn. *Voyager* photographed its rings, discovered several tiny new moons, and showed the moon Miranda to be in a class by itself — Miranda's scarred landscape most probably resulted from a catastrophic collision with a huge asteroid millions of years ago.

Incredibly, *Voyager 2*, with an estimated design life of just five years, found its way to Uranus's twin Neptune, arriving exactly on time on August 24, 1989. It was 12 years and 4 days since the probe had lifted off from Cape Canaveral. Thanks to painstaking reprogramming of its computer software, the images *Voyager 2* sent back were some of the finest ever. They revealed a deep blue, cloud-flecked planet, with more than a passing resemblence to Earth. Again, it was a moon — Triton —

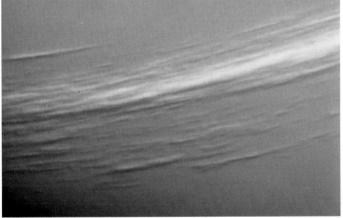

Saturn and satellites
This montage of *Voyager 1* images of the ringed planet Saturn is dominated by its glorious rings, which exhibit mysterious dark "spokes". The moons are, moving clockwise from the right: Tethys, Mimas, Enceladus, and, in the foreground, Dione. Above are Rhea and orange Titan, the only moon known to have a thick atmosphere.

Man on Mars
It will be a very long time indeed before human beings follow in the footsteps of Voyager and visit the gas giants. They are far too cold and inhospitable, and in any case they have no solid surface to walk on. The inner planets Mercury and Venus are also unlikely to be visited in the forseeable future because they are much too hot. Only Mars is a suitable target for future human exploration. It has a solid surface, a slight atmosphere, and temperatures that occasionally rise above freezing point. Astronauts wearing spacesuits would be quite at home there.

This planet has held a particular fascination for Earth people ever since the Italian astronomer Giovanni Schiaparelli reported seeing "canali" on its surface in 1877. This word, meaning channels, was mistranslated as canals, implying artificial waterways made by intelligent beings. Accordingly, many people a century ago believed that there was a race of Martians building canals to channel precious water from the polar ice caps to farming regions near the equator! Space probes have found no evidence of canals or anything remotely artificial on the surface.

Exploring the Red Planet
Both NASA and Soviet space planners are planning to send a manned mission to Mars early next century, and maybe in a spirit of cosmic *glasnost* they will go together. The problems surrounding such a trip are formidable because of the great distance involved — a round trip of 100 million miles (160 million kilometers) or more. New technologies will be needed, particularly in relation to propulsion. Nuclear power could be the answer, provided the crew could be suitably shielded.

The launch window for a manned flight to Mars would need to be chosen carefully to reduce transit time. There are particularly favorable launch windows for flyby missions in the years 2001 and 2016. A round trip would take about 15 months. At other times, and for landing missions, the round trip could take two years or more. Staying weightless for such a length of time might not be too much of a problem, provided the crew excercised regularly. The main problem would be life support, because only limited amounts of food, oxygen, and other necessities could be carried. Careful recycling to conserve resources would be essential.

Neptune's atmosphere
In August, 1989, *Voyager 2* returns this remarkable picture of Neptune from a distance of about 98,000 miles (157,000 kilometers). It shows bands of high cloud casting shadows on the lower cloud deck.

that upstaged the planet during the encounter. Triton has the lowest temperature yet recorded in the solar system, −393°F (−236°C). Astoundingly, volcanoes and geysers dot its surface! But they eject liquid gas, probably nitrogen, rather than molten rock.

At Neptune, *Voyager 2* staged a magnificent finale to the greatest unmanned journey of exploration there has ever been. As it began heading for the stars, Voyager mission scientist Laurence Soderblom commented: "What a way to leave the solar system!"

◀◀
Mission to Mars
Two Viking spacecraft go into orbit around Mars in the summer of 1976. Here one is releasing its lander, still cocooned in its protective aeroshell. Later the aeroshell will separate and the lander will descend to the surface, slowed down to a gentle landing, first by a parachute and then by retrorockets.

◀
Viking lander
Sitting on the Martian surface, the Viking lander is a highly sophisticated scientific station. The arm to the right carries meteorology sensors to monitor the Martian weather. The extendable arm in the foreground collects samples of soil to place in the lander's automatic laboratory. The two turret-like structures are cameras.

▼
Martian panorama
A 360-degree look around the Martian landscape at Chryse from the Viking lander. The horizon appears much closer than it does on Earth because Mars is a much smaller planet. Note the Bicentennial Fourth of July emblem at right. At left is the arm carrying the meteorology sensors.

Clouds on Mount Olympus
In the early morning, wisps of cloud veil the lower slopes of Mars's highest peak, Olympus Mons (Mount Olympus), in a picture based on Viking images. This extinct volcano, the largest in the solar system, towers 15 miles (25 kilometers) above the surrounding landscape.

The summit crater
Great lava cliffs circle the summit crater of Olympus Mons, which measures some 40 miles (65 kilometers) across. This picture shows an artist's model of the crater based on images sent back by the *Mariner 9* probe.

▶

Lava flows
A Viking image of Olympus Mons, processed in false color to show up the successive lava flows that have occurred over many millions of years.

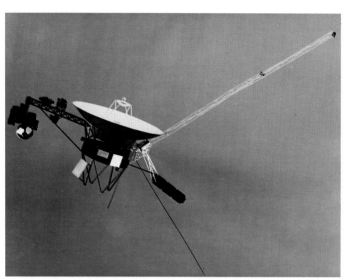

The Voyager probe
A dish antenna 12 feet (3.7 meters) across is the most dominant feature of Voyager. At right is the 43-foot (13-meter) boom that carries magnetometers to measure planetary magnetic fields. At upper left is the science boom, which carries most of Voyager's instruments. At bottom right is the set of three RTGs, nuclear devices that produce electricity to power the instruments.

▶

The Jovian atmosphere
The atmosphere of Jupiter is as colorful as a painter's palette. This *Voyager 2* image shows the turbulence associated with the Great Red Spot, which is about 17,000 miles (28,000 kilometers) long and 9,000 miles (14,000 kilometers) wide.

Pioneer over Jupiter
The Pioneer probes blaze a trail to the giant planets for the Voyagers to follow. *Pioneer 10,* launched in March, 1972, survives the dangers of the asteroid belt to encounter Jupiter in December, 1973. This picture shows *Pioneer 10* speeding past the Great Red Spot.

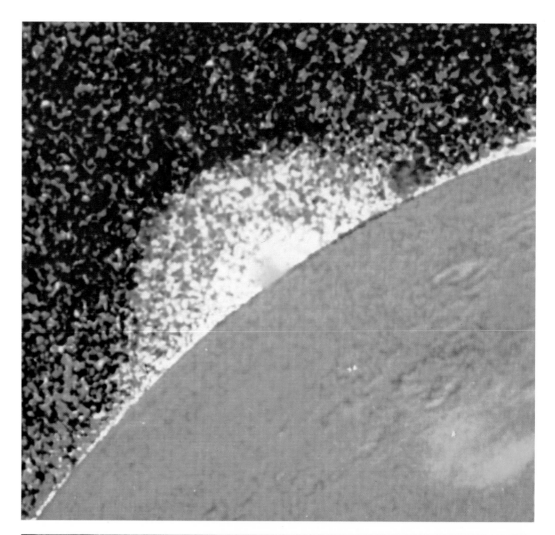

Volcano on Io
Io provides one of the biggest
surprises of the Voyager
encounters with Jupiter.
Intricate processing of imaging
data shows it to be volcanically
active. In this false-color view,
a volcano is erupting on the
moon's limb. It is ejecting a
cloud of fine dust 125 miles
(200 kilometers) above the
surface.

Groovy Ganymede
Ganymede is Jupiter's and the
solar system's largest moon,
3,278 miles (5,276 kilometers)
in diameter. It is made up of a
mixture of rock and ice. This
close-up *Voyager 2* picture of
the moon displays a curious
grooved terrain and bright icy
craters.

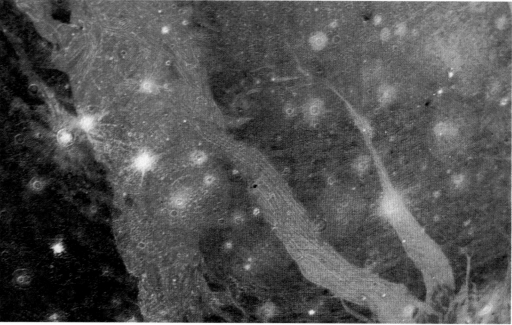

***Galileo* to Jupiter**
The Jupiter probe *Galileo* sets
out from Earth in October
1989 to follow a circuitous
route to the giant planet via
Venus, Earth, and the
asteroids. In the fall of 1995,
Galileo will approach Jupiter
and release a probe into its
thick atmosphere to report on
the conditions there.

Through the ring plane
Saturn is regarded as the jewel of the solar system because of the beautiful rings that circle its equator. The rings are made up of rocky particles of all shapes and sizes, as this 1972 pre-probe artwork portrays.

▶ Glorious rings

Saturn's rings have an overall diameter of 170,000 miles (270,000 kilometers) and would completely dwarf the Earth. They are made up of hundreds of ringlets of rocky particles of different size and composition. False colors have been used in this image to exaggerate these differences.

"View from Mimas"

Saturn rises above the limb of Mimas in this painting by Ron Miller. The moon was almost shattered by a huge meteorite long ago. The collison left an enormous impact crater, seen in the foreground, measuring 80 miles (130 kilometers) across.

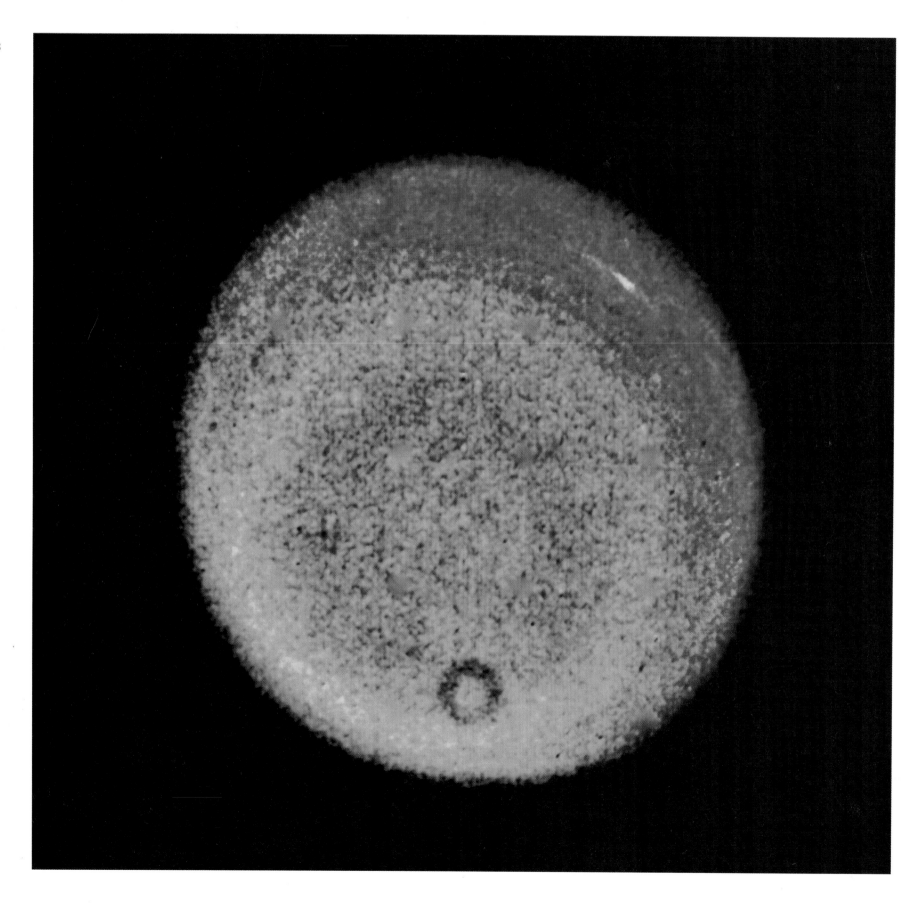

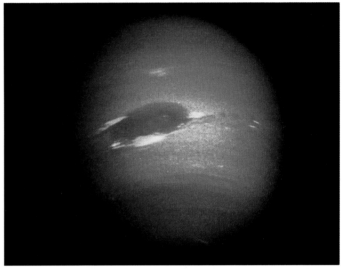

Uranus from Miranda
From the tiny moon Miranda, about 300 miles (485 kilometers) across, Uranus would loom large, as in this montage of *Voyager 2* images. The rings have been added by an artist for effect, but in reality they would be too faint to be noticed from Mimas.

▲
The Great Dark Spot
Jupiter has its Great Red Spot, Neptune its Great Dark Spot, seen in this *Voyager 2* image acquired in August, 1989. It is a false-color image taken through ultraviolet, violet, and green filters. The Great Dark Spot, which is girdled by puffs of cloud, appears to be the eye of a huge storm.

◀◀
Bland Uranus
In natural color the topsy-turvy planet Uranus appears a uniform blue-green. Even extended computer-processing of the imaging data sent back by *Voyager 2* brings out little detail other than a tiny cloud at top right. (The circles are caused by dust in the camera optics.)

Labyrinth of the Night
Just after sunrise, two
geologists explore the
spectacular canyon region of
Mars known as Noctis
Labyrinthus, the Labyrinth of
the Night. Viking pictures have
shown that the canyons are
coated with frost and filled with
mist in the early morning.

Dust storm

Two Mars explorers are caught out by one of the frequent dust storms that the winds whip up on Mars. Their spacesuits have added protection to prevent puncturing, but for safety they must return to their vehicle.

Man on Mars

The year is 2040 and the manned exploration of Mars has been underway for more than a decade. The Mars base has been set up on the equator near an extinct volcano, Pavonis Mons. It includes habitation modules, greenhouses, a launch and landing facility, and a water-well drilling rig and pumping station. Overland transport is provided by a pressurized truck-trailer, while a specially designed airplane exploits the faint Martian atmosphere.

Manned mission chronology

1961
***Freedom 7* — May 5**
Crew: Alan B. Shepard
Duration: 15 min 28 sec

***Liberty Bell 7* — July 21**
Crew: Virgil I. Grissom
Duration: 15 min 37 sec

1962
***Friendship 7* — February 20**
Crew: John H. Glenn
Duration: 4 hr 55 min

***Aurora 7* — May 24**
Crew: M. Scott Carpenter
Duration: 4 hr 56 min

***Sigma 7* — October 3**
Crew: Walter M. Schirra
Duration: 9 hr 13 min

1963
***Faith 7* — May 15**
Crew: L. Gordon Cooper
Duration: 1 day 10.3 hr

1964
***Gemini 3* — March 23**
Crew: Virgil I. Grissom
John W. Young
Duration: 4 hr 53 min

***Gemini 4* — June 3**
Crew: James A. McDivitt
Edward H. White
Duration: 4 days 1.9 hr

***Gemini 5* — August 21**
Crew: L. Gordon Cooper
Charles Conrad
Duration: 7 days 22.9 hr

***Gemini 7* — December 4**
Crew: Frank Borman
James A. Lovell
Duration: 13 days 18.6 hr

***Gemini 6* — December 15**
Crew: Walter M. Schirra
Thomas P. Stafford
Duration: 1 day 1.9 hr

1966
***Gemini 8* — March 16**
Crew: Neil A. Armstrong
David R. Scott
Duration: 10 hr 41 min

***Gemini 9* — June 3**
Crew: Thomas P. Stafford
Eugene A. Cernan
Duration: 3 days 20 min

***Gemini 10* — July 18**
Crew: John W. Young
Michael Collins
Duration: 2 days 22.8 hr

***Gemini 11* — September 12**
Crew: Charles Conrad
Richard F. Gordon
Duration: 2 days 23.3 hr

***Gemini 12* — November 11**
Crew: James A. Lovell
Edwin E. Aldrin
Duration: 3 days 22.6 hr

1968
***Apollo 7* — October 11**
Crew: Walter M. Schirra
Donn F. Eisele
Walter Cunningham
Duration: 10 days 20.1 hr

***Apollo 8* — December 21**
Crew: Frank Borman
James A. Lovell
William A. Anders
Duration: 6 days 3 hr

1969
***Apollo 9* — March 3**
Callsigns: CSM *Gumdrop*
LM *Spider*
Crew: James A. McDivitt
David R. Scott
Russell L. Schweickart
Duration: 10 days 1 hr

***Apollo 10* — May 18**
Callsigns: CSM *Charlie Brown*
LM *Snoopy*
Crew: Thomas P. Stafford
John W. Young
Eugene A. Cernan
Duration: 8 days 3 min

***Apollo 11* — July 16**
Callsigns: CSM *Columbia*
LM *Eagle*
Crew: Neil A. Armstrong
Michael Collins
Edwin E. Aldrin
Duration: 8 days 3.3 hr

***Apollo 12* — November 14**
Callsigns: CSM *Yankee Clipper*
LM *Intrepid*
Crew: Charles Conrad
Richard F. Gordon
Alan L. Bean
Duration: 10 days 4.6 hr

1970
***Apollo 13* — April 11**
Callsigns: CSM *Odyssey*
LM *Aquarius*
Crew: James A. Lovell
John L. Swigert
Fred W. Haise
Duration: 5 days 22.9 hr

1971
***Apollo 14* — January 31**
Callsigns: CSM *Kitty Hawk*
LM *Antares*
Crew: Alan B. Shepard
Stuart A. Roosa
Edgar D. Mitchell
Duration: 9 days 2 min

***Apollo 15* — July 26**
Callsigns: CSM *Endeavour*
LM *Falcon*
Crew: David R. Scott
Alfred M. Worden
James B. Irwin
Duration: 12 days 7 hr

1972
***Apollo 16* — April 16**
Callsigns: CSM *Casper*
LM *Orion*
Crew: John W. Young
Thomas K. Mattingly
Charles M. Duke
Duration: 11 days 1 hr

***Apollo 17* — December 7**
Callsigns: CSM *America*
LM *Challenger*
Crew: Eugene A. Cernan
Ronald E. Evans
Harrison H. Schmitt
Duration: 12 days 13 hr

1973
***Skylab 2* — May 25**
Crew: Charles Conrad
Joseph P. Kerwin
Paul J. Weitz
Duration: 28 days 49 min

***Skylab 3* — July 28**
Crew: Alan L. Bean
Owen K. Garriott
Jack R. Lousma
Duration: 59 days 11 hr

***Skylab 4* — November 16**
Crew: Gerald P. Carr
Edward G. Gibson
William R. Pogue
Duration: 84 days 1 hr

1975
ASTP — July 15
Crew: Thomas P. Stafford
Vance D. Brand
Donald K. Slayton
Duration: 9 days 1 hr

1981
STS-1 — April 12
Orbiter: *Columbia*
Crew: John W. Young
Robert L. Crippen
Duration: 2 days 6 hr

STS-2 — November 12
Orbiter: *Columbia*
Crew: Joseph H. Engle
Richard H. Truly
Duration: 2 days 6 hr

1982
STS-3 — March 22
Orbiter: *Columbia*
Crew: Jack R. Lousma
Charles G. Fullerton
Duration: 8 days 4 min

STS-4 — June 27
Orbiter: *Columbia*
Crew: Thomas K. Mattingly
Henry W. Hartsfield
Duration: 7 days 1 hr

STS-5 — November 11
Orbiter: *Columbia*
Crew: Vance D. Brand
Robert F. Overmyer
William B. Lenoir
Joseph P. Allen
Duration: 5 days 2 hr

1983
STS-6 — April 4
Orbiter: *Challenger*
Crew: Paul J. Weitz
Karol J. Bobko
F. Story Musgrave
Donald H. Peterson
Duration: 5 days 24 min

STS-7 — June 18
Orbiter: Challenger
Crew: Robert L. Crippen
    Frederick C. Hauck
    John M. Fabian
    Sally K. Ride
    Norman E. Thagard
Duration: 6 days 2 hr

STS-8 — August 30
Orbiter: Challenger
Crew: Richard H. Truly
    Daniel C. Brandenstein
    Dale A. Gardner
    Guion S. Bluford
    William E. Thornton
Duration: 6 days 1 hr

STS-9, *Spacelab 1* — November 28
Orbiter: Columbia
Crew: John W. Young
    Brewster H. Shaw
    Owen K. Garriott
    Robert A. R. Parker
    Ulf Merbold
    Byron K. Lichtenberg
Duration: 10 days 7 hr

1984
STS 41-B — February 3
Orbiter: Challenger
Crew: Vance D. Brand
    Robert L. Gibson
    Bruce McCandless
    Ronald E. McNair
    Robert L. Stewart
Duration: 7 days 23 hr

STS 41-C — April 6
Orbiter: Challenger
Crew: Robert L. Crippen
    Francis R. Scobee
    Terry J. Hart
    George D. Nelson
    James D. A. van Hoften
Duration: 6 days 23 hr

STS 41-D — August 30
Orbiter: Discovery
Crew: Henry W. Hartsfield
    Michael L. Coats
    Steven A. Hawley
    Judith A. Resnik
    R. Michael Mullane
    Charles D. Walker
Duration: 6 days 56 min

STS 41-G — October 5
Orbiter: Challenger
Crew: Robert L. Crippen
    Jon A. McBride
    Sally K. Ride
    Kathryn D. Sullivan
    David C. Leestma
    Paul Scully-Power
    Marc Garneau
Duration: 8 days 5 hr

STS 51-A — November 8
Orbiter: Discovery
Crew: Frederick H. Hauck
    David M. Walker
    Dale A. Gardner
    Joseph P. Allen
    Anna L. Fisher
Duration: 7 days 23 hr

1985
STS 51-C — January 24
Orbiter: Discovery
Crew: Thomas K. Mattingly
    Loren J. Shriver
    Ellison S. Onizuka
    James F. Buchli
    Gary Payton
Duration: 3 days 33 min

STS 51-D — April 12
Orbiter: Discovery
Crew: Karol J. Bobko
    Donald E. Williams
    S. David Griggs
    Jeffrey A. Hoffman
    M. Rhea Seddon
    Edwin Garn
    Charles D. Walker
Duration: 6 days 23 hr

STS 51-B, *Spacelab 3* — April 29
Orbiter: Challenger
Crew: Robert F. Overmyer
    Frederick D. Gregory
    Norman E. Thagard
    William E. Thornton
    Don L. Lind
    Lodewijk van den Berg
    Taylor G. Wang
Duration: 7 days 8 min

STS 51-G — June 17
Orbiter: Discovery
Crew: John O. Creighton
    John M. Fabian
    Shannon W. Lucid
    Steven R. Nagel
    Patrick Baudry
    Prince Sultan A. A. Al-Saud
Duration: 7 days 1 hr

STS 51-F, *Spacelab 2* — July 29
Orbiter: Challenger
Crew: Charles G. Fullerton
    Roy D. Bridges
    Anthony W. England
    Karl G. Henize
    F. Story Musgrave
    Loren W. Acton
    John-David Bartoe
Duration: 7 days 22 hr

STS 51-I — August 27
Orbiter: Discovery
Crew: Joseph H. Engle
    Richard O. Covey
    William F. Fisher
    John M. Lounge
    James D. A. van Hoften
Duration: 7 days 2 hr

STS 51-J — October 3
Orbiter: Atlantis
Crew: Karol J. Bobko
    Ronald J. Grabe
    David C. Hilmers
    Robert L. Stewart
    William Pailes
Duration: 4 days 1 hr

STS 61-A, *Spacelab D1* — October 30
Orbiter: Challenger
Crew: Henry W. Hartsfield
    Steven R. Nagel
    Guion S. Bluford
    James F. Buchli
    Bonnie J. Dunbar
    Reinhard Furrer
    Ernst W. Messerschmid
    Wubbo Ockels
Duration: 7 days 44 min

STS 61-B — November 27
Orbiter: Atlantis
Crew: Brewster H. Shaw
    Bryan D. O'Connor
    Mary L. Cleave
    Sherwood C. Spring
    Jerry L. Ross
    Rodolfo N. Vela
    Charles D. Walker
Duration: 6 days 21 hr

1986
STS 61-C — January 12
Orbiter: Columbia
Crew: Robert L. Gibson
    Charles F. Bolden
    Steven A. Hawley
    George D. Nelson
    Franklin R. Chang-Diaz
    Robert J. Cenker
    C. William Nelson
Duration: 6 days 2 hr

STS 51-L — January 28
Orbiter: Challenger
Crew: Francis R. Scobee
    Michael J. Smith
    Judith A. Resnik
    Ronald E. McNair
    Ellison S. Onizuka
    Gregory B. Jarvis
    S. Christa McAuliffe
Duration: 73 sec

1988
STS-26 — September 29
Orbiter: Discovery
Crew: Frederick H. Hauck
    Richard O. Covey
    John M. Lounge
    David C. Hilmers
    George D. Nelson
Duration: 4 days 2 hr

STS-27 — December 2
Orbiter: Atlantis
Crew: Robert L. Gibson
    Guy S. Gardner
    Jerry L. Ross
    R. Michael Mullane
    William M. Shepherd
Duration: 4 days 9 hr

1989
STS-29 — March 13
Orbiter: Discovery
Crew: Michael L. Coats
    John E. Blaha
    James P. Bagian
    James F. Buchli
    Robert C. Springer
Duration: 4 days 23.6 hr

STS-30 — May 4
Orbiter: Atlantis
Crew: David M. Walker
    Ronald J. Grabe
    Mary L. Cleave
    Mark C. Lee
    Norman E. Thagard
Duration: 4 days 1 hr

STS-28 — August 8
Orbiter: Columbia
Crew: Brewster Shaw
    Richard N. Richards
    James C. Adamson
    David C. Leestma
    Mark N. Brown
Duration: 5 days 1 hr

STS-34 — October 18
Orbiter: Atlantis
Crew: Donald E. Williams
    Michael J. McCulley
    Ellen S. Baker
    Franklin R. Chang-Diaz
    Shannon W. Lucid
Duration: 4 days 23.7 hr

STS-33 — November 22
Orbiter: Discovery
Crew: Frederick D. Gregory
    John E. Blaha
    Manley L. Carter
    F. Story Musgrave
    Kathryn C. Thornton
Duration: 5 days 7.5 min

1990
STS-32 — January 9
Orbiter: Columbia
Crew: Daniel C. Brandenstein
    James D. Wetherbee
    Bonnie J. Dunbar
    Marsha S. Ivins
    David D. Low
Duration: 10 days 21 hr

STS-36 — February 28
Orbiter: Atlantis
Crew: John O. Creighton
    John H. Casper
    David Hilmers
    R. Michael Mullane
    Pierre J. Thuot
Duration: 4 days 10.3 hr

STS-31 — April 24
Orbiter: Discovery
Crew: Loren J. Shriver
    Charles F. Bolden
    Bruce McCandless
    Kathryn D. Sullivan
    Steven A. Hawley
Duration: 5 days 1 hr

Glossary

ablation
The melting and boiling away of a heat shield during re-entry, a process that dissipates the heat produced by air friction.

abort
Cut short a flight or a mission.

acquisition
Making contact with a spacecraft so that signals can be transmitted to it or received from it.

aimpoint
The point in space that a space probe is aimed at so as to make its intended planetary or satellite encounter.

airlock
A chamber in a spacecraft which can be pressurized with air and depressurized, through which astronauts leave and enter their craft when they go spacewalking.

ALSEP
The Apollo lunar surface experiments package, which formed the basis of the automatic scientific stations the Apollo astronauts set up on the Moon.

apogee
The most distant point a satellite reaches in its orbit around the Earth.

apogee motor
A rocket fitted to a spacecraft which is fired at the high point of an elliptical parking orbit to place it in geostationary orbit.

artificial gravity
A force similar to the force of gravity, created in a space station, say, by rotating the station on its axis.

artificial satellite
A man-made object put into orbit around the Earth, or other planet or moon. Usually just termed satellite, though strictly "satellite" refers to a natural body, such as the Earth's Moon.

ASTP
The Apollo-Soyuz Test Project, the joint flight in July 1975 of Apollo and Soyuz astronauts and cosmonauts.

asteroids
Minor planets or small rocky bodies in the solar system that circle the Sun mainly in a broad band (the asteroid belt) between the orbits of Mars and Jupiter.

astronautics
The science of space travel; hence astronaut, a space traveller.

atmosphere
The layer of gases surrounding a planet or a moon; the Earth's atmosphere is composed mainly of nitrogen and oxygen. But the atmospheres of other planets are composed of a totally different gas mixture — that of Venus, for example, is mainly carbon dioxide.

attitude
The position of a craft in relation to something else, such as the horizon.

aurora
A display of colored lights and streamers seen mainly in polar regions, caused by particles in the solar wind interacting with gases in the upper atmosphere. Auroras have also been detected on other planets.

back-up
A person or item of equipment available to take over the function of another person or item should the need arise. For many space missions there are prime and back-up crews.

blackout
The loss in communications with a spacecraft that is re-entering the atmosphere, caused by ionization of the surrounding air. Medically a blackout is loss of consciousness that can occur when a person undergoes high g-forces.

booster
The first stage of a launch vehicle or additional rocket stages attached to the core vehicle to give extra thrust at lift-off.

burn
The period during which a rocket fires.

CapCom
Abbreviation for Capsule Communicator, the person at Mission Control who communicates with astronauts in space.

capsule
The name given to the cramped crew cabin of early spacecraft such as Mercury and Vostok.

celestial
Relating to the heavens.

celestial mechanics
Study of the mechanisms governing the motions of heavenly bodies, satellites, and probes.

centrifuge
A training aid for astronauts, a machine that whirls them round in a capsule at the end of a long arm, subjecting them to centrifugal forces that duplicate the g-forces they will experience on lift-off and re-entry.

cosmic rays
Charged particles that come from outer space, mainly protons and electrons.

cosmos
An alternative term for the universe; space.

countdown
The counting down to zero of a certain period of time before a space launch.

CSM
The combined command and service modules of the Apollo spacecraft.

de-orbit
Come out of orbit; usually effected by a de-orbit burn, or a firing of retrorockets.

deploy
Place in position.

docking
The joining together of two spacecraft in space.

drag
The resistance experienced by a body traveling through air.

DSN
NASA's Deep Space Network, a communications and tracking network for distant space probes operated by Jet Propulsion Laboratory in California.

EMU
Extravehicular mobility unit, the correct term for the Space Shuttle spacesuit.

encounter
A meeting between a space probe and its target.

ESA
The European Space Agency, the body responsible for coordinating space activities in Europe.

escape velocity
The speed a body must possess to escape completely from Earth's gravity; about 25,000 mph (40,000 km/h).

EVA
Extravehicular activity, activity outside a spacecraft; commonly called spacewalking.

expendable
Can be used only once, as in expendable launch vehicle.

extraterrestrial
Existing outside the Earth.

false-color image
One in which the colors are not true to life.

flyby
A space mission in which a probe flies past a planet or moon without going into orbit around it or landing.

free fall
The state that exists in orbit when everything is "falling around the Earth". It gives rise to the phenomenon we know better as weightlessness.

fuel cell
An electric cell that produces electricity by combining hydrogen and oxygen gases to form water; used to supply electrical power in the Apollo spacecraft and the Space Shuttle.

g
The symbol for the acceleration due to gravity. We can say the force on our bodies due to gravity is 1g; see **g-forces**.

gantry
A tower that gives access to a launch vehicle on the launch pad.

geostationary orbit
An orbit 22,300 miles (35,900 kilometers) high, in which a satellite circles the Earth once every 24 hours and therefore appears to be fixed in the sky.

getaway special
An experiment carried in a small container in the Space Shuttle. The getaway special program gives universities and other organizations access to space at reasonable cost.

g-forces
The forces a body experiences when it is subjected to high acceleration or deceleration, e.g. during a rocket lift-off or spacecraft re-entry; 3g, for example, means that the forces experienced are three times the normal pull of gravity.

glitch
A problem of some sort.

global positioning system
A U.S. satellite navigation system for shipping, consisting of 18 Navstar satellites.

gravity-assist
A method that uses the gravitational attraction of a planet to increase the speed of a space probe.

g-suit
Clothing worn by astronauts and pilots that applies pressure on the lower body and legs to prevent blood collecting there under high g-forces.

hatch
A door in the hull of a spacecraft with an airtight seal.

heat shield
A coating on the outside of a spacecraft designed to protect it and any astronauts inside from the heat developed during re-entry. Both the U.S. and Soviet Shuttles use ceramic tiles as a heat shield.

hold
A temporary halt in a countdown.

ignition
The instant when a rocket engine is started, when the propellants are ignited. Thereafter the propellants continue to burn by themselves.

Intelsat
The International Telecommunications Satellite Organization.

interplanetary
Between the planets.

interstellar
Between the stars.

jettison
Discard. Parts of a spacecraft may be jettisoned before re-entry.

launch vehicle
A system of rockets that can propel a payload into space; a step rocket.

launch window
The period of time during which a spacecraft can be launched so that it can reach its planned target or objective.

life-support system
The system in a spacecraft that keeps the astronaut crew alive and in a comfortable environment.

light-year
The distance light travels in a year, nearly 6 million million miles (10 million million kilometers), which provides a useful unit of measurement in astronomy.

LM
Sometimes LEM; the Apollo lunar excursion module.

LOX
Liquid oxygen.

lunar
Relating to the Moon.

Mach number
The speed of a craft through the air compared with the local speed of sound.

magnetosphere
The great "bubble" that extends way out into space around a heavenly body with a magnetic field where the solar wind interacts with the magnetic field. The magnetosphere is irregular in shape, rather like a raindrop, with the blunt end facing into the solar wind.

manned maneuvering unit
See **MMU.**

mare
A lunar sea; plural maria.

meteorite
Rock from outer space that has survived entry through the atmosphere and fallen to the ground.

meteoroid
A particle of dust or rocky matter traveling in space, seen as a meteor when it hits the Earth's atmosphere.

mission
A space flight.

mission control
The communications and control center for a space mission. The U.S. mission control center for manned space flight is located at the Johnson Space Center, Houston, Texas. U.S. satellite mission control is located at the Goddard Space Flight Center, Greenbelt, Maryland. Mission control for planetary explorations is at the Jet Propulsion Laboratory, Pasadena, California.

mission specialist
A Shuttle astronaut with specific duties related to a particular flight, especially with regard to the payload.

MMU
Manned maneuvering unit; a jet-propelled backback used by Shuttle astronauts for moving about in space.

mock-up
A full-size dummy spacecraft that looks like the real thing, often fitted with instruments and controls, which may be "live". See also **simulator.**

module
A major self-contained unit of a spacecraft.

moonquake
Ground tremors in the Moon's crust, caused by internal activity or the impact of meteorites.

multistage rocket
Another name for a step rocket.

NASA
National Aeronautics and Space Administration; the body that organizes aerospace activities in the United States.

navsat
A navigation satellite.

NOAA
National Oceanic and Atmospheric Administration; the organization that manages the U.S. weather and Earth-resources satellite programs.

nominal
A term which means that everything is as it should be.

NORAD
The North American Air-Defense Command, which constantly monitors all satellites (and debris) in orbit.

OMS
Orbital maneuvering system of the Shuttle; two engines that fire to inject the vehicle into orbit and to bring about de-orbit.

OMV
Orbital maneuvering vehicle; a robot rocket vehicle used to move satellites and other units in orbit; being developed for use with the space station *Freedom.*

orbit
The path in space taken by one body traveling around another. Orbits are usually elliptical.

orbital maneuvering system
See **OMS.**

orbital maneuvering vehicle
See **OMV.**

orbital period
The time it takes a satellite to complete one orbit. For a satellite orbiting a few hundred miles above the Earth, the orbital period is about 1½ hours.

orbital velocity
The speed a satellite needs to remain in orbit. Around the Earth at an altitude of about 200 miles (300 km), the orbital velocity is about 17,500 mph (28,000 km/h).

orbiter
A spacecraft designed to orbit a planet or moon; specifically the Space Shuttle orbiter.

outer planets
The planets Jupiter, Saturn, Uranus, Neptune, and Pluto.

oxidizer
The propellant in a rocket that provides the oxygen to burn the fuel.

PAM
Payload assist module; a booster rocket stage attached to a payload to boost it into a high orbit; used on the Space Shuttle and the Delta rocket.

payload
The cargo a launch vehicle or rocket carries.

payload specialist
A Shuttle astronaut who goes into space, probably only once, to supervise the operation of a certain experimental payload.

perigee
The point in a satellite's orbit when it is closest to the Earth.

plasma
A gas cloud in which the gas exists as a mixture of ions.

PLSS
Portable life-support system; the self-contained backback astronauts wear when they go spacewalking.

probe
A spacecraft that escapes from the Earth's gravity and travels to the Moon or the planets.

propellant
A substance burned in a rocket engine to develop thrust to propel it. Two propellants are generally needed, a fuel (such as liquid hydrogen or kerosene) and an oxidizer (such as liquid oxygen).

radiation belts
Intense regions of radiation around planets that have a magnetic field, caused by the presence of highly energetic charged particles.

radio telescope
A telescope, usually with a large metal dish, designed to gather radio waves from the heavens.

RCS
Reaction control system; a system of thrusters on a spacecraft fired to change its attitude in space.

redundancy
The duplication of vital parts in a system, so that if one fails, another can take over.

re-entry
The moment when a spacecraft or vehicle hits the atmosphere when returning from space. The drag, or resistance of the air, slows the craft down but causes it to heat up due to air friction.

remote sensing
The gathering of data and imagery from a distance; especially of the Earth's surface from satellites, such as Landsat.

rendezvous
The meeting of two craft, e.g. in orbit prior to docking.

resolution
The ability to distinguish details in, say, a photograph or image.

retrofire
Firing a rocket backwards, i.e. in the direction of travel, to act as a brake; hence retrobraking.

RMS
Remote manipulator system of the Shuttle; the Shuttle's "crane", a remote-controlled jointed arm fitted into the orbiter's payload bay.

rocket
An engine that produces a stream of hot gases for propulsion, working on the reaction principle. It can work in airless space because it carries not only fuel, but also the oxygen to burn the fuel.

RTG
Short for radioisotope thermoelectric generator, a nuclear power source that converts the heat produced by the decay of a radioactive isotope into electricity to power a spacecraft.

satellite
A small body that orbits around a larger one in space, such as the Moon around the Earth. The Moon is Earth's natural satellite. However, the term satellite these days invariably means artificial satellite, a man-made moon.

sarsat
A search and rescue satellite, equipped to relay signals from emergency beacons on board planes and ships.

scan platform
The movable part of the Voyager spacecraft, which carries instruments that need to be accurately pointed.

shepherd moon
One that orbits near a planetary ring and helps to keep the ring particles in check.

shirt-sleeve environment
One in which people can wear ordinary clothing, with no need for pressure suits or spacesuits.

simulator
A full-scale mock-up of a spacecraft which simulates the behavior of the real craft.

solar
Relating to the Sun.

solar cell
A silicon-wafer device that produces electricity when sunlight falls on it.

solar system
The Sun and the family of planets, moons, asteroids, meteors, and comets that circle around it.

solar wind
The stream of charged particles emitted into space by the Sun.

space colony
A habitat built in space that could provide a permanent home for thousands of people.

space medicine
The study of the human body in space, particularly with regard to the effects on it of a weightless environment.

space sickness
The nausea experienced by many astronauts during the early part of a space mission. The condition is properly called space adaptation syndrome.

spacesuit
A multilayer garment astronauts wear in space as protection against the vacuum, cold, and harmful radiations.

spacewalk
The popular name for EVA (extravehicular activity).

speed of light
In a vacuum light travels at a speed of about 186,000 miles (300,000 km) per second. Radio waves and all other electromagnetic waves also travel at this speed, which, according to relativity theory, is the highest speed at which anything can travel.

splashdown
The landing of a spacecraft at sea.

SRB
Solid rocket booster; two are used on the Shuttle.

stage
One of the rocket sections of a step rocket.

stellar
Relating to the stars.

step rocket
A rocket made up of several rocket units (stages) joined together, usually end to end.

STS
Space Transportation System; the correct name for the U.S. Space Shuttle system. Shuttle flights are designated by an STS number.

suborbital
A space flight in which a craft does not go into orbit, but follows a ballistic trajectory.

TCM
Trajectory correction maneuver.

TDRS
Tracking and data relay satellite; a powerful communications satellite in geostationary orbit used to relay tracking information, communications, and data from low-flying satellites, particularly the Space Shuttle, into NASA's communications network.

telemetry
Transmitting instrument readings between a spacecraft and ground control.

terrestrial
Relating to the Earth.

terrestrial planets
Rocky planets like the Earth — Mercury, Venus, and Mars.

thrust
The pushing force that propels a rocket, measured in newtons (SI units) or pounds.

thruster
A small rocket engine used for maneuvering.

time-lag
The finite time it takes for radio signals to travel between a spacecraft and the Earth.

touchdown
The moment of landing of a spacecraft.

tracking
Following the path of a spacecraft through space by radio, radar, or photography.

tracking and data relay satellite
See **TDRS.**

trajectory
The flight path of a body.

trajectory correction maneuver
Firing thruster rockets on a spacecraft so as to change its speed, and therefore its trajectory, through space.

umbilical
A connecting line carrying power, air, water, and oxygen between, say, a launch-pad gantry and a launch vehicle; or between an astronaut and his or her spacecraft.

VAB
Vehicle Assembly Building; the massive building at Kennedy Launch Complex 39, built for the assembly of Saturn V Moon rockets and now used for assembling the Space Shuttle.

weightlessness
The condition astronauts experience in orbit, when their bodies appear not to have any weight. The proper term for this state is free fall.

zero-g
Zero-gravity; another term for weightlessness.

Entries in quotation marks refer to names of paintings; page references in italics refer to captions to illustrations.

Acknowledgements/Credits

The author would like to extend his grateful thanks for their unstinting advice and encouragement to his many friends at NASA establishments around the country, and particularly at NASA Headquarters in Washington DC. The majority of the photographs illustrating this book have been provided by NASA, through Spacecharts Photo Library, to whom many thanks.